WHISPERS OF BREATH

The Art and Science of Spirituality

NAINA AND Dr. ROHIT SHETTY

STARDOM BOOKS

www.StardomBooks.com

STARDOM BOOKS, LLC
112, Bordeaux Ct,
Coppell, TX 75019

FIRST EDITION JUNE 2023

STARDOM BOOKS

A Division of Stardom Alliance
112, Bordeaux Ct,
Coppell, TX 75019

www.stardombooks.com

Stardom Books, United States
Stardom Books, India

WHISPERS OF BREATH

Naina and Dr. Rohit Shetty

p. 160
cm. 12.7 X 20.32
Category:
OCC027000 - Body, Mind & Spirit : Spiritualism - General
OCC011020 - Body, Mind & Spirit : Healing - Prayer &
Spiritual

ISBN: 978-1-957456-21-8

DEDICATION

"To Dad,
your wisdom guides us".

Dedicated to everyone on a spiritual journey and those seeking to be on this path. Our sole intention is to spread the understanding of our very TRUE nature.

"We have been born to experience every moment and to live our lives to the fullest…"

CONTENTS

FOREWORD

The aim of every soul, for that matter, every living being, is to find peace and happiness. I'm so overjoyed that Rohit and Naina have come out with a book, giving us guidelines on how we could bring peace and contentment into our lives. Another brilliant aspect of this book is that not only does it take a spiritual route to explain the meaning of contentment, but it also teaches us how to accept things as they are and exemplifies taking responsibility for what, where, and how we are, without blaming others, for the things that are happening to us, is the secret to happiness and peace. Another great secret they have brought out elegantly is the "attitude of gratitude." Their presentation of the idea of gratitude demonstrates that millions of things are going right in our lives. We simply need to focus on them and be grateful for them than to crib about the few things that aren't going right.

I wish all the readers of this book all health and happiness in their lives.

Dr. K Bhujang Shetty
Founder
Narayana Nethralaya Eye Hospital

A NOTE FROM NAINA SHETTY

This book could not have been moved into the world without the help and support of many. First, I wish to acknowledge my father, Dr.K.Bhujang Shetty, who has profoundly impacted my inner world with deep gratitude. This book is his inner voice; I am only a channel that brought it out into these pages through these words. I have very vivid memories of him in my childhood; he would patiently tell me stories pertaining to life and its various perspectives. As we traveled through this life, he was always present to guide me through real-time experiences that I faced alone or as a family and the many interactions we had with the outside world. His energy runs through me and pushes me to 'seek' the truth.

I also have a special dedication to my mother, Rajkamal Shetty, who I genuinely look up to in awe of her 'grace' and her 'inner beauty' that shines through us as a family. I truly only reflect her light. I'd also like to acknowledge my children, Iksha and Ishan, who are truly the reason for my spiritual growth. I acknowledge my brother, Naren, sister-in-law, Namratha, in-laws Seetharama, Hemalatha, and Radhika, and to all my friends and well-wishers who have genuinely supported and stood by my side through all the peaks and troughs of my inner journey.

I deeply thank my mentor, Himalayan Siddhaa Akshar, who has instilled divine teachings in my Yogic journey. My humble appreciation for the teachings in Pranic healing which was instilled in me at the right moment in my life.

And finally, thank you to the co-author of this book, my

husband, Dr. Rohit, who has motivated me to pen my inner experiences.

The Origin

Every experience, small and big, in my life has been my Guru. When I look back now as I write these words and ask myself, "How did I turn into 'Me'?" I move past multiple flashbacks of memories profoundly imprinted in my subconscious mind. These imprints from my childhood have been very special moments that I have shared with my father.

I have been in awe of him since I was a little girl. I look up to him and love his 'inner beauty.' I observed him very closely as a little girl from the time when he would empathetically converse with his patients. He could imagine how the other person was feeling.

What are emotions?

How do we decide how to react to a particular experience or situation, or person?

Through my experiences, I have realized that 'my emotions' is the programming of my subconscious mind.

My mother would be very busy with her regular chores; asking me to accompany my father to his clinic. Today I'm very glad and grateful that she made this decision.

I used to be seated on a small chair at the corner of his consultation room, and all I did was "observe." It never ceased to amaze me how a stranger would walk into the room with fear in their eyes and would walk back with a smile reflected in their eyes. I realized then that it was just not the examination of the eyes that they were pleased with. It was mainly the power of a few words of empathy that instilled in them an unquestionable faith in my father.

As an observant child, this was my first learning. A lesson learned on "the power of energies instilled with faith and hope." The significant transformation in the energy of their eyes from fear stricken to complete surrender was definitely due to my father's smile, which never drooped, and was given to every patient who walked in. This smile was always on his face throughout a long, exhausting day. I'd never known then how this observation would profoundly impact my life.

The consultation room experiences with my Dad were and are my most precious imprint in me.

I was receptive to what emotions were all about and how they influenced me, and how they reflected the immediate world outside. Even today, when I am low on energy, this incident inspires me to "smile" and uplift my energies.

ACKNOWLEDGMENTS

Through the eyes of gratitude, everything is a miracle.
- Mary Davis

The creation of this book was a labor of love, and it is with the greatest gratitude that we acknowledge the efforts of our little team of doctors on their spiritual journey that helped breathe life into its pages.

To Reshma Ranade and Bhavya Gorimanipalli – we are thankful for the meticulous review of literature spanning decades that brought us scientific research and evidence-based medicine to understand the impact of our practices on a biological level.

To Priyanka Sathe Inamdar and Isha Acharya – we owe a debt of gratitude for using their remarkable artistic talent to create beautiful illustrations and eye-catching sketches that ease comprehension and make reading an engaging experience.

And we thank Maithri Arunkumar and Sailie Shirodkar – for using their linguistic skills to put pen to paper – for scripting these ideas into words that could make their way into the minds and hearts of many with ease. Finally, we thank the Universe for supporting us in our endeavors to make the journey toward holistic well-being an enjoyable one– for all.

Wishing you the very best on your journey,

Naina and Dr. Rohit Shetty.

PREFACE

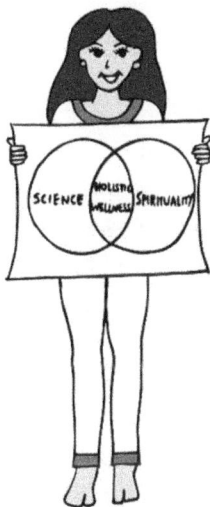

"Science is not only compatible with spirituality; it is a profound source of spirituality."

— Carl Sagan

Spirituality and science share a complex history. In traditional settings, the same person often dispensed medical and spiritual care. However, technological advancements and emerging sciences, have witnessed a paradigm shift in our approach to healthcare and living. It has been aptly observed that the focus of modern medicine has shifted from being a caring and service-oriented model to a technological and cure-oriented model[1]. While the latter has helped achieve dramatic progress in cure rates and recoveries – **would a complete detachment from the former truly be in our best interests?**

"Helping, fixing, and serving represent three different ways of seeing life. When you help, you see life as weak. When you fix, you see life as broken. When you serve, you see life as whole. Fixing and helping may be the work of the ego, and service the work of the soul."
— Rachel Remen, MD

Dualism in Healthcare

'Dualism' in healthcare – or the practice of separating spiritual and physical well-being – has been a historically important landmark in that it permitted medical practice to be divorced from religious oversight. The formal separation of the 'mind' from the 'body' granted dominion over the incorporeal mind to religion and spirituality, while medical science related itself entirely to the body.[2] This stark separation of mental and physical health has governed our medical practice.

However, it has also made our practice more reductionist and dispassionate and promoted a problematic view of complementary and alternative medicine. This is probably one of the reasons why patients tend to lose a sense of self when they enter hospitals – names and personalities are often reduced to hospital IDs, complex diagnoses, and other minutiae that diminish their individualities compared to their identity as a patient and an illness.

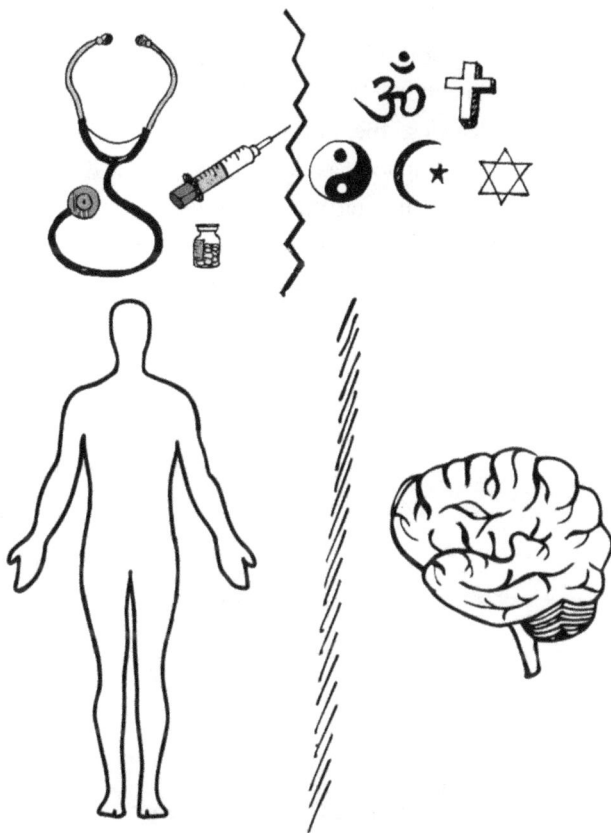

The stark separation of mental from physical health has governed our medical practice.

By introducing spirituality into our healthcare practices and daily lifestyles, we may humanize these otherwise barren experiences.

What is Holistic Wellness?

"The part can never be well until the whole is well."
— Plato

The perception of 'dualism' has pervaded not only our practice of healthcare but also our daily lifestyles. However, in recent decades, the notion of holistic healthcare has gained more traction as we have grappled with the possibility that mental and physical health may be inextricably linked. Meditation and mindfulness techniques are known to be practices that deliver mental health benefits in the form of reduced anxiety, stress, and depression[3]. Yet, they have also recently been demonstrated to have clinical benefits in those suffering from physical ailments such as hypertension, insomnia, irritable bowel syndrome, etc.[4]

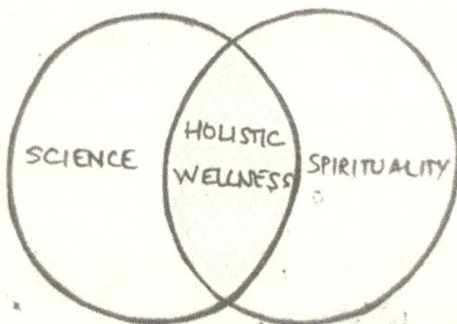

For the sceptics amongst us, the idea of spiritual health is still a notion that is difficult to come to terms with. However, as the old adage goes, "there are no atheists in foxholes." Studies have been conducted to demonstrate the relationship between awareness of terminal illnesses and spiritual well-being - it has been noted that terminally ill patients tend to have increased spirituality over their non-terminally ill or healthy counterparts.[5,6] Research has demonstrated that patients living with chronic illness are often attracted to complementary medicine and the practice of self-care.[7] It is often through these life-altering experiences that we attempt to seek a higher meaning to our existence. Our goal, through these pages, is to promote holistic well-being as a way of living rather than an afterthought – to have it seep into the essentials of our routine living and become an indispensable aspect of our routines. We aim to do this by eradicating this notion of 'duality' – the physical and the mental, the tangible and the abstract – and offer a more comprehensive view of holistic self-care practices with peer-reviewed scientific literature.

Our team of spiritual teachers and practitioners of Western medicine have collaborated to put together practices to enrich your lifestyles, equipped with the knowledge of how it works. We present this to you with the sincerest hope that it assists you in transcending physical, mental, and spiritual barriers.

Would You Like to Read Further?

1. Puchalski CM. The role of spirituality in health care. Proc (Bayl Univ Med Cent). 2001;14(4):352-357.

2. Gendle MH. The Problem of Dualism in Modern Western Medicine. Mens Sana Monogr. 2016;14(1):141-151.

3. Cramer H, Hall H, Leach M, et al. Prevalence, patterns, and predictors of meditation use among US adults: A nationally representative survey. Sci Rep. 2016;6:36760.

4. Ando M, Morita T, Akechi T, et al. The efficacy of mindfulness-based meditation therapy on anxiety, depression, and spirituality in Japanese patients with cancer. J Palliat Med. 2009;12(12):1091- 1094.

5. Leung KK, Chiu TY, Chen CY. The influence of awareness of terminal condition on spiritual well-being in terminal cancer patients. J Pain Symptom Manage. 2006;31(5):449-456.

6. Paloutzian, R. F., & Ellison, C. W. (1982). Loneliness, Spiritual Well-Being, and the Quality of Life. In L. A. Peplau, & D. Perlman (Eds.), Loneliness: A Sourcebook of Current Theory, Research and Therapy (pp. 224-236). New York: John Wiley & Sons.

7. Adams, J., Kroll, T., & Broom, A. (2014). The significance of complementary and alternative medicine (CAM) as self-care: Examining 'hidden' health-seeking behavior for chronic illness in later life. Adv Integr Med, 1(3), 103-104.

1

Art and Science of Spiritual Awareness

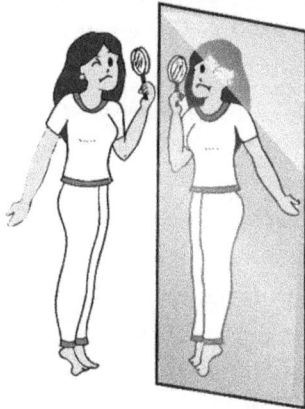

Mirror, mirror on the wall,
Who's the most aware of them all?
"Awareness is like the sun. When it shines on things, they are transformed." — Thich Nhat Hanh

Awareness is a word that is used to signify consciousness. To be aware is to be conscious.

In the daily grind of life, our living has become mechanical. We function on autopilot mode. We have a set routine of habitual patterns that we follow automatically. Such is the automated nature of our duties that even our response to experiences has become mechanical.

Our experience has been so dictated by these external mechanisms that we have adopted these qualities and called them a part of our personalities.

For example, if someone responds in anger to a certain situation in their life, that person will then come to describe themselves as angry.

But are you, though?

If you want to gain Awareness in your lives, you need to learn to disassociate the emotion from yourself. You are not the emotion. Chant this mantra to yourself: "Anger is an emotion moved into me. I'm not an angry person. And I am not the emotion itself."

The recognition of this separation is vital. When you recognize this fact, you have taken your first and foremost step toward Awareness.

Why is this recognition important? Once you realize the difference between emotions and yourself, you can easily detach yourself from the trappings of these emotions. If you don't realize this difference, these emotions will become a part of the fabric of your personality, and you will find it harder to untangle yourself.

Now you might be confronted by the question: what defines our personality?

Our personality or character is defined most importantly in the first seven years of our life. The activities and habits you imbibe during this period become like programs you feed into your subconscious mind. This programming develops your character.

Now, **what are these programs?**

We develop certain programs through our experiences with a situation or a relationship and feed them into our subconscious minds. It could be habitual behavioral patterns, how we perceive emotions like love, joy, sorrow, and hate, and how we respond to multiple behavioral patterns.

So, you have to ask yourself whether you want to evolve into a better being, or are you satisfied with what you are now and simply claim, "This is who I am"?

If you do wish to change, you are ready to take the next step. This step involves your resolve and willingness to evolve. Only when you are sure of your will to evolve can you begin the process of reprogramming your mind.

Enquire yourself with these questions: What are your first thoughts in the morning? Are they the same habitual thoughts? Are you still plagued with stress about your incomplete tasks? Are you anxious about how to complete those tasks? Or are you stricken with doubt about your ability to complete those tasks? Your thoughts most natural tendencies might be about a list of tasks to be completed to ensure your livelihood or pondering on a negative experience that you have encountered the day before.

The emotions that you feel in the morning are the most powerful. These emotions you feel during this time move directly into your subconscious mind, and you can successfully reprogram your mind.

Why is it important to have your subconscious mind be affected for you to reprogram and become aware?

Only 5% of our lives do we live consciously; the rest, 95% of our life, we live through the programming we created during our initial years. When I first heard this, I was awed by this concept of the energy-mind-body. Now you may understand why it is so hard to develop a new habit or cull an existing negative habit. Reprogramming takes a lot of willpower, perseverance, and most importantly, the repetition of particular habits.

So let us return to the point when I said early morning thoughts are crucial and the key to your mind's reprogramming. Just think briefly about how you are when you have just woken up.

The pangs of sleep can be felt in your drooping eyelids even as the twitches of your body remind you that you are awake. This period is the intersection of your conscious mind and subconscious mind. Both these states of being are merged at this moment. This merging is vital for your reprogramming because this is the moment when you are undisturbed by the filter of your ingrained habits, which will block you from moving into your subconscious realm. Trying to reprogram your mind at any other time of the day and you might find yourself riddled with unnecessary and distracting thoughts calling into question the effectiveness of the new patterns or habits you are trying to cultivate.

Thus, early morning is the ideal time when you can feed all that you want to manifest, your dreams, your goals, and the changes that you wish to see in your life into your subconscious mind. Think in terms of which emotion you would want to work on. What relationship would you like to heal in your life? This process can be understood as conscious awareness.

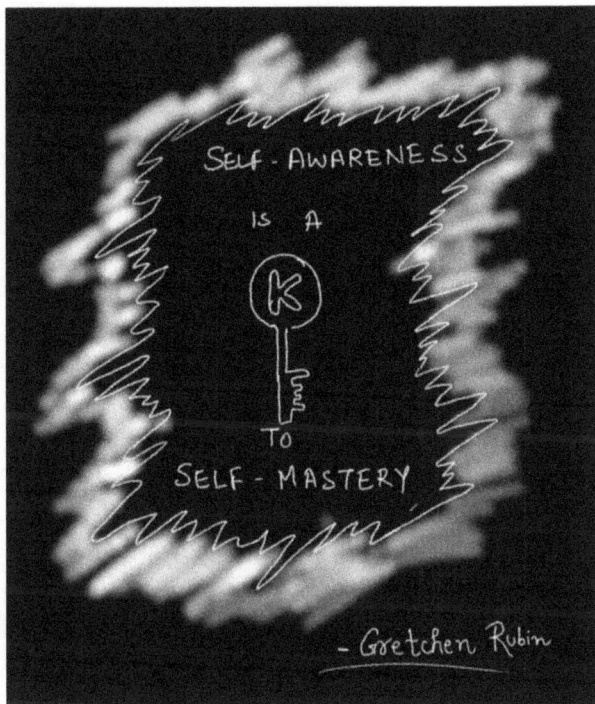

SELF - AWARENESS

IS A

To

SELF - MASTERY

— Gretchen Rubin

Methodology

1. **Perception of your day**: Just like how important your thought process is when you get up from bed, the time when you retire to bed to fall asleep is equally essential. This moment is very precious. Be extremely aware of your thoughts during these times. Be a witness to your thoughts. Allow only positive thoughts to flow as a conscious decision. Even if you have had a bad day, it is vital that you do not reflect on the day negatively. Instead, reflect on your day with all the challenges you faced and look at the positive side of the experiences. Whatever has to happen will happen, none of us can

avoid that. The only control that we have in our life is how we perceive our life and situations. While we live in this world, we exist and thrive mostly in our minds. So, make sure to keep your mind the most cleansed and beautiful place to live in.

2. **Affirmations**: This is another crucial habit you should look to cultivate. Once you have looked back on your day and reflected on the day after accepting all the experiences of the day, be it positive or negative, you need to think about the changes you would want to bring into your life. Observe the state where you are about to fall asleep. This is the time when your filters do not function due to the merging of your conscious and subconscious minds. This is another phase of reprogramming. Repeat affirmations of the changes that you want to see in your life. For example, if you feel someone has wronged you or you have wronged someone else, the affirmations would be, "I forgive so and so" or "I ask forgiveness from such and such person/s."

3. **Visualization**: The next important step we need to do is to build your visualization. How? Let us continue with the same example. Let us say your relationship with someone has been damaged due to your or the other person's mistake. Visualize in your mind that the relationship has been healed.

   How would you describe the facial expressions of you and your friend? Observe the faces very clearly in every vivid detail. Try and focus on how the bliss and peace are reflected in the faces. Observe the calmness and feel the love of the relationship return. Bask in that warmth and you will soon find that your mind is also healing.

4. **Experiencing the bliss of healing**: Feel the bliss of how you would feel when you have healed this relationship. So when you visualize the scene from the previous example, look to find the resonance in your reality. To really heal, the crucial step is to vibrate at the same frequency of the emotion we felt in our visualization of the healing of the relationship. Feel that bliss and feel the healing. Feel this awareness when you fall asleep.

The Four 'R's of Being Aware

1. REFLECT on your day with positivity
2. REPEAT affirmations of the changes you want in life
3. REFOCUS on happier mental images
4. REALIZE the joy of healing

Are Awareness, Visualisation, and Willpower a Science?

Consciousness is a mysterious yet essential fact of life. We are born without it, but we develop consciousness over the first few years of our lives. The neuronal network of human consciousness appears to be simpler than the other higher cortical functions such as vision[1].

As Luder Deecke puts it, "Consciousness is like a platform or stage or screen in the brain; it is like the executive floor of a company, where only agendas ready to be signed are let up, but the groundwork has already been done by the unconscious pre-processing. So, consciousness is the highest 'authority,' so to say."[2]

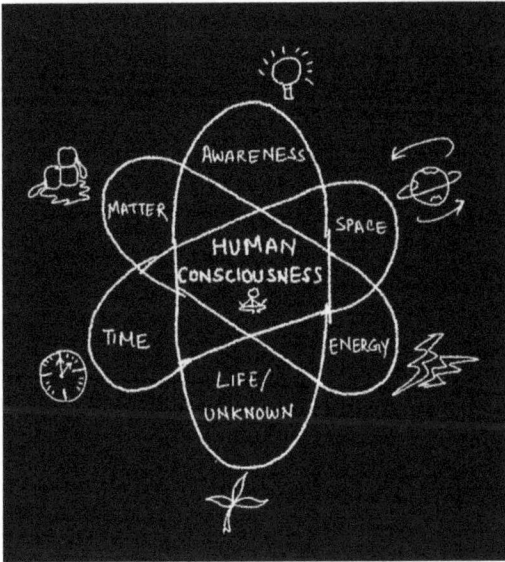

Awareness is defined as the ability to selectively direct attention to distinct aspects of the environment and to be able to manipulate these aspects of the environment over a prolonged duration than usual cognitive processing will allow[1].

Self-awareness, a trait that is hard to acquire but not impossible, has also shown benefits in one's workplace. It alleviates stress and allows the development of psychologically healthy humans. Self-awareness is said to have different components, similar to what is described by our spiritual leaders – reflection, rumination, mindfulness, and insight.

A self-reflection and insight scale that was developed by Anthony Grant and John Franklin helped people to assess themselves in a new light.[3] Similarly, a self-awareness outcomes questionnaire (SAOQ)[4] was designed to know the effects of self-awareness and mindfulness in individuals.

The questionnaire had simple one-line determinants such as –"I am content with my work situation", "I am aware of my abilities and limitations", "I feel emotions deeply", "I can 'take a step back' from situations to understand them better".

They were tested on a group of 215 individuals, from students to professionals, who were also made to engage in six self-awareness practices such as meditation, prayer, mindfulness practice, talking therapy, writing a journal, and participating in a personal development group. They found that those who reflect more are more likely to experience self-development outcomes, and they were also shown to be more proactive at work. The individuals who engaged in self-awareness practices were found to correlate with increased acceptance of themselves.

The Power of Visualisation

Mental imagery or visualization is the key to improving mental health. There are two components to it – mental imagery (MI), which is imagining a place or imagining oneself being at a place, and mental practice (MP), which means visualizing mentally a task one wants to do, say, lift weights or sing at a concert. Many famous artists or sports persons visualize themselves performing well before they actually go up on stage or onto the field. It is said that MP and MI boost overall performance by focusing attention and preventing disruptive or negative thoughts.[5] Visualization activates the same areas in the brain as the physical activity itself and stimulates the visual cortex as well. There have been many experiments conducted, and studies performed to demonstrate the benefits of visualization. PET scans have shown that the blood flow to areas responsible for certain muscular activities is increased just by visualization. For example, if a person was asked to only visualize walking out the door to familiar surroundings and moving around – the parietal cortex of the brain was shown to receive increased regional blood flow and 10% more oxygen.[6]

In fact, a study demonstrated that just by visualizing the little finger moving in abduction (outward), a 35% improvement was seen in the abducting capacity of the little finger.[7] A study conducted among 72 participants with assigned tasks in their daily lives demonstrated that the group to whom motivational imagery was shown completed more tasks compared to the control groups. They also showed higher levels of motivation to complete the tasks.[8] If motivational imageries could help a group of people complete given tasks, imagine what we could do by visualizing our achievements every day!

There is also symbolic mental imagery, e.g., imagining a bird balancing on one leg on a beam can symbolize the desired outcome, such as being able to balance oneself on a beam. Such forms of visualization are used by healthcare professionals, sports coaches, art teachers, etc., to not just motivate a patient or student but to improve performance clinically and in real life.

Physical therapists use the modality of mental imagery and mental practice to help patients regain motor functions, especially those who have been immobile for a while.[9] The quiet and concentration required to perform mental practice give the additional advantage of de-stressing a patient. It allows patients to visualize themselves performing physical movements in real-life situations, instilling hope and the motivation to recover.

It takes awareness, self-consciousness, and willpower to be able to visualize and work toward achieving the goals one has set.

"Visualization is more important than knowledge"
— Albert Einstein

The Famous Marshmallow Experiment

Way back in 1972, Walter Mischel, a psychologist at Stanford University, decided to conduct an experiment[10] using items that would test a child's willpower to its limits – food! Marshmallows, pretzels, or cookies were kept on a table in front of 32 preschool children, and they were told that they could eat what was kept on the table, but if they waited for a period of 15 minutes, they would get a second treat. The room was free of distractions, and some children gave in to the temptation and ate the sweets while some did not. This experiment laid the foundation for a series of experiments on "delayed gratification," which the scientists felt was a method to assess the willpower of an individual.

Why Delay Gratification?

"As we get past our superficial material wants and instant gratification, we connect to a deeper part of ourselves as well as to others, and the universe."

— Judith Wright

Are you wondering what delayed gratification has anything to do with awareness? Let me explain – It takes willpower to not give into temptation, and by not giving in, we allow ourselves to experience a greater sense of satisfaction when we are rewarded.

To be able to have willpower, we need to be aware of the situation in front of us, our present needs and wants, our goals for the future, and the consequences of giving in to the said temptation. It allows us to make better and more informed decisions throughout our lives and that is exactly what the follow-up studies of the Marshmallow study proved.

The Neurobiology of Delayed Gratification

If you thought that was the end of the marshmallow study, let us fast forward 40 years – all those pre-schoolers are now 40-something adults going about their lives, and a repeat experiment was conducted on them. This time around, their brains were studied in depth using a functional MRI scan, and guess what?

They proved that there are functional differences in the brains of those who delayed gratification versus those who did not. An important area in our brain called the prefrontal cortex contributes to impulse control, predicts the consequences of an action, and governs decision-making to a great extent.

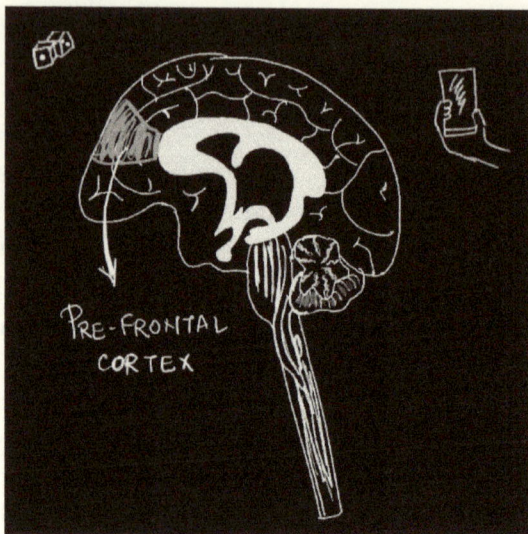

Pre-Frontal Cortex

The follow-up studies 11-13 of the marshmallow experiment showed that those adults who, as children, were able to have the willpower to wait and then eat the treats were performing better through their adolescent and adult lives, while those who were less able to delay gratification as children displayed low self-control in their twenties and adult lives. The fMRIs supported evidence by demonstrating diminished activity in the right prefrontal cortex in those individuals who had given into temptation as children.

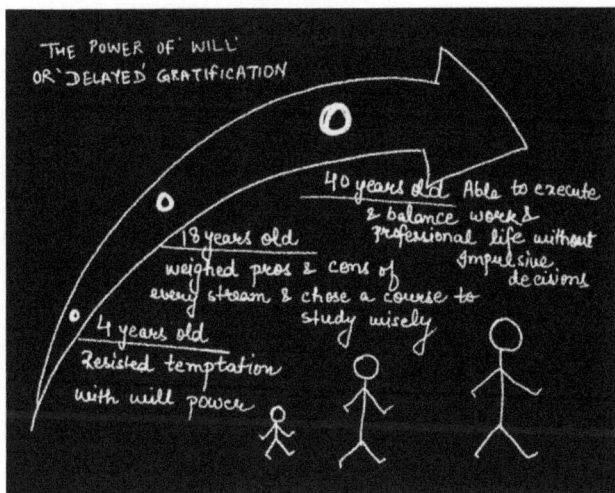

THE POWER OF WILL OR DELAYED GRATIFICATION

40 years old Able to execute & balance work & professional life without impulsive decisions

18 years old weighed pros & cons of every stream & chose a course to study wisely

4 years old Resisted temptation with will power

They also found that those who had waited longer in this situation at the age of four were described by their parents as more academically and socially competent as compared to their peers a decade later and more able to handle frustrating situations and resist temptation. They were noted at statistically significant levels to be more verbally fluent and able to express ideas clearly. They were found to have qualities such as attentiveness and increased concentration and were noted to be more self-assured. This difference was also reflected in their Scholastic Aptitude Test (SAT) scores[10]. This emphasizes the need to imbibe this psychological process and particularly encourage it to be taught to children at a young age.

So you see, to be aware is an art, and awareness is a science in itself. Spiritual awareness seeks to make you a more wholesome human being to live well-rounded, full lives.

Awareness is like the Sun. When it shines on things, they are transformed. ~ Thich Nhat Hanh

Anthony Grant
John Franklin
↓
Self-reflection
& Insight
Scale

SPIRITUAL AWARENESS

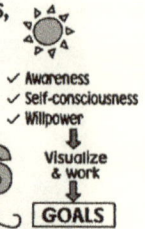

✓ Awareness
✓ Self-consciousness
✓ Willpower
↓
Visualize
& work
↓
GOALS

◆ Dissociate
◆ You≠Emotion
◆ Recognize
◆ Positive Affirmations
◆ Visualize

• Rumination
• Mindfulness
• Insight

● REFLECT
● REPEAT
● REFOCUS
● REALIZE

To be AWARE is
To be CONSCIOUS ...

Would You Like to Read Further?

1. Charlton B. Evolution and the cognitive neuroscience of awareness, consciousness, and language. Cognition. 2000; 50:7–15.
2. 2. Deecke L. There are conscious and unconscious agendas in the brain and both are important-our will can be conscious as well as unconscious. Brain Sci. 2012;2(3):405-20.
3. Grant, A. M., Franklin, J., & Langford, P. (2002). The Self Reflection and Insight Scale: A new measure of private self-consciousness. Social Behavior and Personality: An International Journal, 30(8), 821–835
4. Sutton A. Measuring the Effects of Self-Awareness: Construction of the Self-Awareness Outcomes Questionnaire. Eur J Psychol. 2016;12(4):645-658.
5. Warner L, McNeill ME. Mental imagery, and its potential for physical therapy. Phys Ther.

1988;68(4):516-21.

6. Roland PE, Eriksson L, Stone-Elander S, et al: Increases of regional cerebral oxidative metabolism and regional cerebral blood flow provoked by visual imagery. Society for Neuroscience Abstracts 12:117,1986

7. Ranganathan VK, Siemionow V, Liu JZ, Sahgal V, Yue GH. From mental power to muscle power-- gaining strength by using the mind. Neuropsychologia. 2004;42(7):944-56.

8. Renner F, Murphy FC, Ji JL, Manly T, Holmes EA. Mental imagery as a "motivational amplifier" to promote activities. Behav Res Ther. 2019; 114:51-59.

9. Fansler CL, Poff CL, Shepard KF: Effects of mental practice on balance in elderly women. Phys Ther 65:1332-1338.

10. W Mischel, Y Shoda, MI Rodriguez, Delay of gratification in children. Science 244, 933–938 (1989)

11. Casey BJ, Somerville LH, Gotlib IH, et al. Behavioral and neural correlates of delay of gratification 40 years later. Proc Natl Acad Sci U S A. 2011;108(36)

12. IM Eigsti, et al., Predicting cognitive control from preschool to late adolescence and young adulthood. Psychol Sci 17, 478–484 (2006)

13. Y Shoda, W Mischel, and PK Peake, predicting adolescent cognitive and self-regulatory competencies from preschooldelay of gratification: Identifying diagnostic conditions. Dev Psychol 26, 978–986 (1990)

2

"Neurons That Fire Together, Wire Together"

"An atomic habit is a little habit that is part of a larger system. Just as atoms are the building blocks of molecules, atomic habits are the building blocks of remarkable results."

- James Clear

As human beings, we are driven by ambition - a drive to be our best selves. Every New Year brings with it a host of new resolutions, a multitude of detailed to-do lists in freshly purchased planners, and a yearning desire to do/be better than the year past. However, it doesn't take long for most of our wishes to fizzle out; for the throngs crowding every gym to dwindle down to the occasional enthusiast, and for the dust to settle in on bookshelves being meticulously cleaned and organized.

Have we paused to wonder why this occurs? Surely it is not that we do not have the will to achieve our goals – **what is it then that distinguishes the few who actually hold on to their aspirations against all odds and set out to achieve what others cannot?** Is it an elusive quality that only a handful are blessed with – **or is it something that can be cultivated by all of us with time and effort?**

In his book, "Atomic Habits," James Clear lays down a step-by-step approach to achieving our goals without losing the spark that first ignited us. An 'atomic habit' may be defined as a small, convenient daily habit that is not only easy to perform but, when practiced consistently, has the incredible power to produce compound growth and progress. Attaining our pre-determined goals contributes greatly to a feeling of contentment and self-sufficiency. With this in mind, we have outlined some of the key practices from this book to help us in our journey toward mental, spiritual, and holistic well-being.

Success is the Product of Daily Habits...

In a famously quoted line from the book, Clear states, "Success is the product of daily habits—not once-in-a-lifetime transformations." It is difficult to accept this idea when the media continuously bombards us with dramatic displays of success. We see an athlete that no one has heard of before shoot to an astonishing level of fame overnight, and we wonder why we still struggle with meeting our daily goals. The problem lies herein – we are far too focused on the results to pay heed to the process that brought them success; we do not see the years, perhaps decades, of turmoil that culminated into them becoming an 'overnight success.'

While a lot of people believe that it is the big and dramatic transformations that lead to success, it is usually the tiny, barely noticeable yet consistent improvements in our daily habits that build a momentous achievement. **This may sound fantastic in theory, but does it actually measure up? Is there a scientific or mathematical basis for the compounding worth of small yet regular practices?** As it happens, the author very beautifully represents the value of infinitesimal atomic habits over time in a graphical format.

1% WORSE EVERY DAY FOR 1 YEAR = 0.99^{365} = 0.03

1% BETTER EVERY DAY FOR 1 YEAR = 1.01^{365} = 37.78

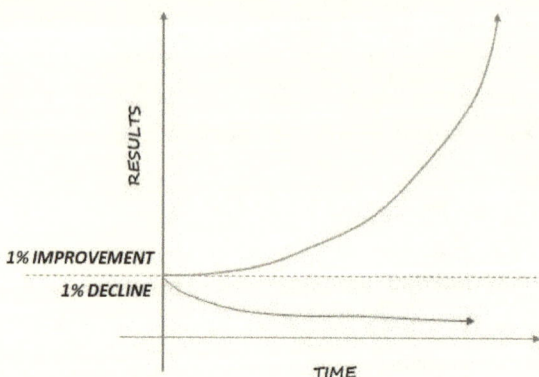

A slight improvement – approximately 1% - in our daily habits can compound to making us 37 times better over a period of one year! It might make one wonder – what is it that you would like to be 37 times better at by the time the year is up? And how do you consistently make marginal improvements to reach that target?

The trick is to shift your focus – from an outcome-based approach to an identity-based approach. Let's make this simpler.

Change your Identity to Transform your Outcomes

The problem with most of our thinking is that we focus far too much on the 'goal' and far too little on the 'process.' How are they different? Goals are the results that we want to achieve, while the process includes definite steps that we take to achieve these. When we fail to form a good habit, it is not the goal of becoming a better person that fails us – it is the process we built to get there. When bad habits repeat themselves, it is not an inherent flaw in our nature – but the process we have picked out, which isn't ideal for the change we aim to create. **So how do we go about building a better system for change?**

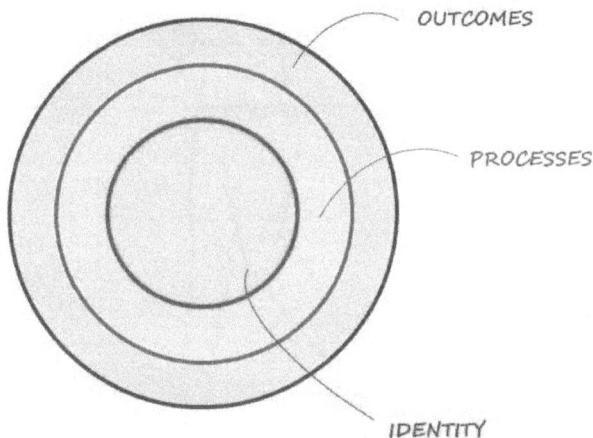

Reproduced from Atomic Habits by James Clear

The key to building habits that will last us a lifetime is to focus on building a new identity that is in alignment with one's goals. Our present behavior mirrors our present identity – the kind of person that we believe we are. In order to change one's behavior, one needs to start believing new things about oneself. How do we do this? First, decide the kind of person you want to be – and then, keep proving it to yourself by achieving small goals. Soon, you will find your improved identity emerging – with every action being a vote in favor of the person you wish to become one day.

"The goal is not to learn an instrument, the goal is to become a musician."

- James Clear

How Do I Build Better Habits?

There are four simple steps to building better habits and making them last– make it *obvious*, make it *attractive*, make it *easy* and make it *satisfying*.

Make it Obvious

The author is of the opinion that it is not a lack of 'motivation' but a lack of 'clarity' that fails us. When one is surrounded by visual cues that prompt one to perform a particular action, it is easier to put it into practice, e.g., placing a bottle of water on your desk is a reminder to hydrate frequently; placing your vitamin supplements in plain sight in the kitchen would make it more likely for you to take them rather than hiding them away in a cabinet. Specifying a time and location to perform a habit also ensures that one is more likely to perform them.

> At 8 am every morning, I will sit at the desk in my study and fill my gratitude journal

Make it Attractive

Habits may be considered dopamine-driven feedback loops. It is the anticipation of a reward that motivates us to act, and this anticipation fuels dopamine surge in the human body. One method to take advantage of this scientific fact is 'temptation bundling.' The author describes this as pairing a habit that you *want* to do with one that you *need* to do. For e.g., rewarding oneself with 15 minutes of your favorite TV show after an hour of studying, getting to play your favorite music while exercising, etc.

Make it Easy

Humans, in general, are notorious for picking the path of least resistance. Picking up a new habit complicated takes a lot of time or requires a lot of effort, and it is less likely that you would stick with it. The manner in which one can make a habit easy is by priming one's environment to support the activity. If the yoga studio is on your way back home from work, it is far more likely that you would make practicing yoga a regular habit. However, if it were to require long walks completely out of the way, it would become much more difficult to make exercising a long-term or, rather, permanent habit.

Make it Satisfying

The first three rules may ensure that an act is performed; the last one ensures that the behavior is repeated. One of the most satisfying aspects of a habit is *progress*. Of course, progress may not always be tangible or measurable, or even instantaneous. **However, the human brain, in general, favors instantaneous rewards over delayed gratification– so how does one facilitate this?** Tracking one's habits is an easily performed daily measure of one's progress – and of course, it serves the mental satisfaction of having stuck to one's goal for one day longer. Identifying such measures of small change makes it that much likelier that one would make a habit stick.

Weekly Habit Tracker

HABIT:	S	M	T	W	T	F	S
_____	☐	☐	☐	☐	☐	☐	☐
_____	☐	☐	☐	☐	☐	☐	☐
_____	☐	☐	☐	☐	☐	☐	☐
_____	☐	☐	☐	☐	☐	☐	☐
_____	☐	☐	☐	☐	☐	☐	☐
_____	☐	☐	☐	☐	☐	☐	☐
_____	☐	☐	☐	☐	☐	☐	☐

Breaking Bad Habits...

The inverse of the four laws described above is a clear path to breaking habits. Make it *invisible* – removing a cue from your environment, such as deleting a time-consuming app from your phone, would make it less likely for you to keep scrolling on social media instead of working. Make it *unattractive* – reminding yourself of the health hazards of smoking would most likely keep you from picking up your next cigarette. Make it *difficult* – increase the friction between you and your bad habit, increase the effort required to perform the action, and you will automatically take the path of least resistance. And lastly, make it *unsatisfying* – make the choices of your bad habits public and painful; having an 'accountability partner' to reprimand you for falling back onto your old habits is one way to do this.

We conclude with the words of the famous neuropsychologist Donald Hebb – "neurons that fire together, wire together." He used this phrase to describe how neural pathways are not only formed but also reinforced through repetition. The more one performs a certain action, the more one reinforces a *habit* and the stronger that neural circuitry becomes. This ensures that the act performed is easier, more efficient, and more successful each successive time it is performed.

Isn't that all the more reason to put these life-changing values into practice – **gratitude, grounding oneself, staying centered, vibrating at a positive frequency?** When these actions become habitual and get ingrained in your systems, it becomes easier to repeat them and reap the rewards along your spiritual journey.

Would you like to read further?

1. Clear, J. (2018). Atomic habits: tiny changes, remarkable results: an easy & proven way to build good habits & break bad ones. New York, New York, Avery, an imprint of Penguin Random House.

3

Power to Stay Centered

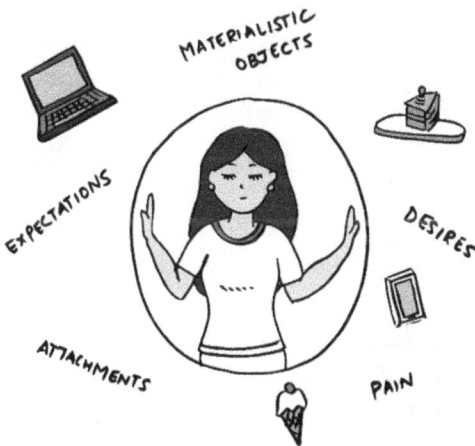

"Attachment is the source of all suffering."
— Buddha

In my spiritual journey, there was a time when I consciously realized that whenever I experienced 'pleasure' of any sort, my mind formed an attachment to this pleasure. This pleasure, in turn, generated cravings which in turn caused pain. When I realized this connection, I understood the term 'attachment'. I realised of why I needed to discard 'attachments' from my life.

When I say discard attachments, I mean to remove the binding emotions of desire, jealousy, hatred, and anger. Let us take, for example, the desire to have vanilla ice cream. If I indulge this desire, I will realize that all objects are changing, and the pleasures they give are very short-lived. The moment I connect pleasure with an object (here being vanilla ice cream), I immediately generate the desire to experience this pleasure. The more I fulfill my desires, the deeper I fall.

When I fall deeper by fulfilling my desires, I will be entangled and enslaved by those desires. It is the same case with pain. The more deeply I feel hurt, the deeper it will leave its imprint. If I fall deeper into the imprint, it would be more difficult for me to come out of it. This was the moment when I realized the importance of cultivating a certain Inner Equipoise. The entire mechanism of extreme likes and dislikes, attachments, and aversions will not affect you in this state. This state of being, in turn, will enhance every aspect of your life personally and professionally.

Let us dive deeper into this with an analogy of water currents. If you observe water currents, you will see that they flow forwards by carving out their grooves on the ground.

Similarly, your thoughts will carve out their own grooves in your mental space. Your thoughts which are the water currents, will leave their imprints in your subconscious mind. These imprints are referred to as *Vasanas* in the *Patanjali Yoga Sutra*. *Vasanas* is a Sanskrit term that refers to a past impression in the mind that influences your present behavior. *Vasanas* can be good or bad. A *vasana* could be made with the desire to eat vanilla ice cream, and another *vasana* could be caused by intense emotional pain. These experiences can cause 'deep grooves' in your subconscious mind, which are so powerful that they push you to experience more of it. So, our mission here is to make sure these 'mental grooves' are not so deep wherein we are not left at the mercy of these currents (thoughts). Our mission is to ensure that we are in control of our minds. Thus, the more non-attached we are, the more we do not bring egoistic inclinations into our lives. It would result in a more natural flow and compassion towards all. We have always confused attachment to love, but we clearly understand that non-attachment is love as it is a selfless act.

Now comes the hard bit of how to sustain these *vasanas*. *Pathanjali Yoga Sutra* tells us that Yoga makes us aware of the mind's impulsive nature. Through regular and sincere Yoga practice, we will be able to consciously develop self-mastery over the mind. This is how we learn to be 'detached'.

In non-attachment, we are unaffected by the world's objects – likes/dislikes/attractions/aversions and consciously stay in a *Samasthithi* state or a balanced state of being.

How do Science and Modern Medicine Define 'Non-Attachment'?

Several religions and cultures have defined the concept

of 'non-attachment', but it is primarily a Buddhist construct. While the Western world likens 'attachment' to affection or love, Buddhist philosophy confers negative connotations to the term. It promotes a non-contingent form of happiness that they label 'non-attachment'.

The term 'non-attachment' has been described to mean engagement with an experience with flexibility and without fixation on achieving specified outcomes. Unlike 'detachment,' which is a passive stance that can be likened to carelessness or indifference, non-attachment allows one's happiness to be free of external influences while still being intimately in touch with one's reality.[1]

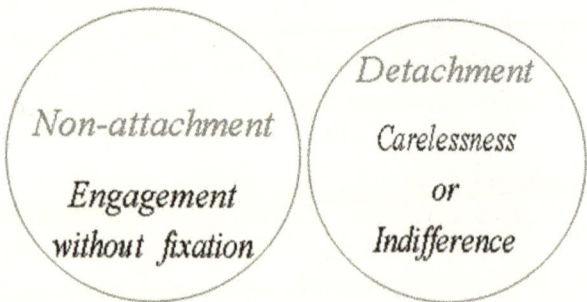

While one can gauge an individual's capacity for non-attachment through traits such as acceptance, the desire to be happy in the face of unfavorable circumstances, autonomy, and security, science has attempted to quantify these through the invention of various scales. Sahdra et al. devised a measure called the Nonattachment Scale (NAS)[2], after consultations with several Buddhist scholars and experts in Buddhist texts. A large sample size of adults and college students helped standardize the inventory, and the resulting measures are structured based on Buddhist principles[3].

Another study investigated the role of nonattachment and mindfulness together using this scale and discovered higher scores on the NAS to be inversely related to suicidal ideation and depression[3]. So you see, the idea of promoting mental health and happiness by being unattached to materialistic possessions, is not merely a flowery virtue propounded over centuries – there is concrete evidence, determined by a validated scale, that expounds on how the extent to which you practice non-attachment, promotes your mental well-being.

Attachments, Addictions, and the Power of Yoga

A dysregulated sense of attachment could potentially lead to more adverse outcomes – addictions, substance abuse, and other related pathologies. It may well be viewed as an extreme form of attachment. The unfettered abuse of different psychoactive drugs is a significant healthcare burden, contributing vastly to psychosocial morbidities, social instability and disability, and occupational crises.

The World Drug Report 2015 asserted that 1 out of 20 people between the ages of 15 and 64 used an illicit drug in 2013, and reported 187,100 drug-related demises in the same year.[4]

"People become attached to their burdens sometimes more than the burdens are attached to them."

— George Bernard Shaw

While the role of modern medicine in treating substance abuse cannot be understated, the importance of meditation as add-on therapy, or even as a primary treatment in an earlier, more benign, non-pathological state of attachment, needs to be realized. Yoga and exercise-based interventions have demonstrated an increase in cognitive function and protection against cognitive decline, which is prevalent among those suffering from substance abuse disorders.[5] Increase in the volume of the bilateral hippocampus, has been demonstrated through a pilot study on yoga intervention in elderly adults. Reversing hippocampal volume loss has been linked to improved memory function, and has been considered a potential mechanism for Yoga-enhanced cognitive function in substance abuse disorders.[6]

"Detachment means letting go and nonattachment means simply letting be."

— Stephen Levine

Mindfulness, which also emphasizes the ability to stay centered, has been inversely related to depression, anxiety, and stress in several studies. It has been suggested as a technique to cultivate non-attachment, which is further linked to lower mental distress. Yoga and mindfulness share a fundamental belief – being aware of experiences and emotions as they arise, without feeling the need to change them. As addictions result from 'mindless' states – escapist attitudes, emotional reactions, and automated thinking – practicing traditional elements of yoga can help steady the mind, enhance concentration, and build emotional resilience[7].

Based on an experimental pre- and post-intervention study of a 10-day Buddhist retreat, an increase in mindfulness led to decreased subjective depression, anxiety, and distress[8]. Yoga and mindfulness-based interventions thus have growing empirical support in enhancing addiction treatment and preventing relapse while promoting recovery.

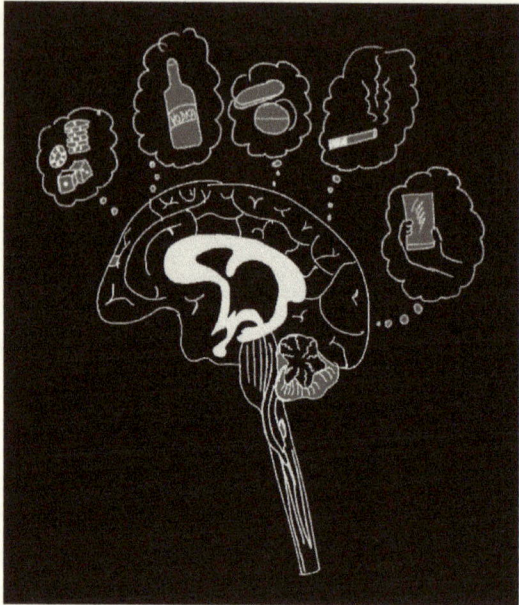

A dysregulated sense of attachment, could potentially lead to more adverse outcomes – addictions, substance abuse and other related pathologies

Addiction is an extremely complicated problem that unfortunately navigates several psychological, social, and biological circumstances, and is often rooted in trauma.

Overcoming any form of addiction is no easy feat – and in extreme, life-threatening situations, the role of immediate medical intervention is unprecedented. However, the use of yoga practices in tandem with medication is often a philosophy that is adopted by several rehabilitation centers for exactly the reasons illustrated above – because while it cannot negate the social and biological causes behind addiction, it can certainly promote our mental capacity to meet these challenges, and can also help one navigate through the emotions of stress, fear, depression, anxiety, and panic accompanying these addictions.

The research on nonattachment is still limited, as the scale to measure it has been very recently developed. However, early studies have demonstrated encouraging results and asserted the role of mindfulness or the power to stay centered in combating psychological distress resulting from a state of attachment.

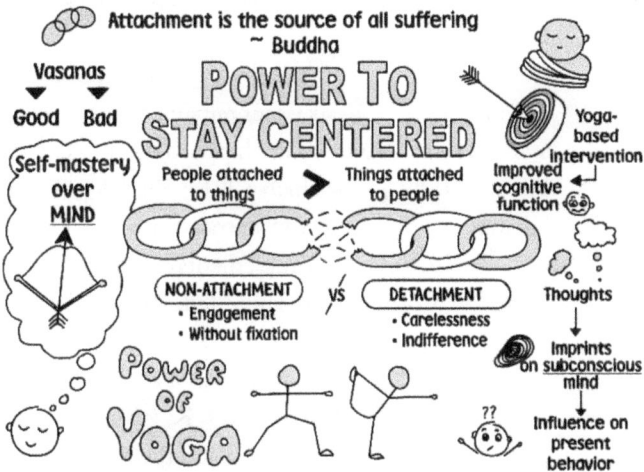

Would You Like to Read Further?

1. Khong, BL. Expanding the Understanding of Mindfulness: Seeing the Tree and the Forest. Humanistic Psychologist. 2009; 37:117– 136.

2. Sahdra, B, Shaver, PR, Brown, K. A scale to measure nonattachment: a Buddhist complement to western research on attachment and adaptive functioning. J Pers Assess. 2010; 92:116–127.

3. Lamis, DA, Dvorak, RD. Mindfulness, nonattachment, and suicide rumination in college students: the mediating role of depressive symptoms [published online October 2013]. Mindfulness. doi:10.1007/s12671-013-0203-0.

4. United Nations Office on Drugs and Crime: World Drug Report 2015. Vol. 2013, United Nations Publication, Sales No. E. 15.XI.6, 2015.

5. Gothe NP, Keswani RK, McAuley E: Yoga practice improves executive function by attenuating stress levels. Biol Psychol 2016; 121: 109–116.

6. Hariprasad V, Varambally S, Shivakumar V, Kalmady SV, Venkatasubramanian G: Yoga increases the volume of the hippocampus in elderly subjects. Indian J Psychiatry 2013; 55: S394–S396.

7. Khanna S, Greeson JM. A narrative review of yoga and mindfulness as complementary therapies for addiction. Complement Ther Med. 2013;21(3):244-252.

8. Ostafin, BD, Chawla, N, Bowen, S, Dillworth, TM, Witkiewitz, K, Marlatt, G. Intensive mindfulness training and the reduction of psychological distress: a preliminary study. Cogn Behav Pract. 2006; 13:191–197.

4

Anger

"For every minute you remain angry, you give up sixty seconds of peace of mind."

— Ralph Waldo Emerson

The quote, *"Speak when you are angry, and you will make the best speech you will ever regret"*, might be familiar to you. Anger is not an unaccustomed emotion, as we all have experienced it. It is a feeling that makes us feel agitated and frustrated towards ourselves and the people around us.

Is your body language during an anger situation similar to your everyday body language? **Have you ever noticed the quick, short breathing pattern you experience while you are engorged with anger?**

We should be consciously aware of our breathing patterns during anger situations. A quick and short breathing pattern will only exaggerate the anger experienced. Anger releases hormones such as adrenaline and cortisol, accelerating heart rate and breathing.

The first and foremost step in tackling anger is to pause. A few seconds of silence and a halt will help save a lot of damage caused by your words or actions. The challenging part of this step is to gain the ability to pause when you encounter an anger situation. A pause clubbed with deep inhalation and exhalation will enable you to maintain a sense of balance to make the right decision with the right words and right actions. Those few seconds are enough to perceive the entire experience that triggered the anger.

The energy associated with anger is thicker and enlarged, and hence it has the power to bind people and situations together. However, we should allow forgiveness to flow from us to mend this energy. This heals tense relations and also helps in showing love and forgiveness with ease.

What exactly is anger? Anger can be easily defined as the non-acceptance of things beyond our control. A readiness to accept things beyond our control is tolerance, which is a much-appreciated value in human beings.

Self-Inquiry

Attaining self-inquiry or spiritual maturity comes with a

long array of changes.

1. Shifting our focus on changing ourselves rather than others and external situations upholds self-inquiry.

2. Readiness to accept people as they are by realizing that everyone is right in their respective perspectives.

3. Working on self-introspection to let go and focusing more on bettering ourselves accelerates the process.

Self-inquiry or self-talk has always been advisable to cut the cord of anger and focus more on ourselves.

Methodology

Step I – Self-Talk:

Self-talk or self-inquiry yields better results if practiced early in the morning. I suggest the period between 3 AM and 5 AM for practicing this technique, as external disturbances will not be an obstacle as they will be null during this time. If you have encountered any anger situation recently, take time to process the situation the next day itself. It enables you to understand whether the way you dealt with the situation was a healthy one or not. Sit on your bed and contemplate the whole experience in your mind. Ask yourselves questions such as, "What action triggered the anger?", "Was it necessary to succumb to anger in that situation?" and so on. Step aside and visualize the whole process from a third person's perspective to properly analyze the entire situation. It is also crucial to check whether we have contemplated all the anger situations we encountered, as it will only help in distracting our anger to some other valuable forms.

Contemplating past events, especially anger situations, is very important. It teaches us how simple adverse reactions tarnish our relationships. You might be unaware of the future consequences it creates as you are stuffed with anger at that moment. Analyzing the experience and healing from it is the most powerful and important step to tackling anger. This step should not be mistaken as an excuse for claiming "life is hard".

Step II – Reprograming:

The second step involves reprogramming your thought process against the person or situation that triggered anger. Instead of cultivating the emotion of hate, try displacing it with forgiveness. Implementing the attitude of forgiveness is not very easy, and hence it is advisable to visualize the process first. Affirmations like "I forgive you" and "I am sorry for hurting you", etc., are indeed powerful statements that make a lot of changes. When practiced in the early morning, these affirmations yield powerful results as you are in the alpha state. Hence, reprogramming occurs very quickly in your subconscious mind. That is why, we say, "when you are at peace within... the world around us is also peaceful".

Step III - Brahmari Pranayama

The third step involves an effective breathing technique called Brahmari Pranayama, which can calm our nervous system instantly when encountering anger. It brings harmony and peace within you. This step disintegrates any residue of the anger component within you even after following the first two steps.

Method

Sit upright and close your eyes. Use both index fingers to close the cartilage of your ears. Inhale and exhale deeply, and while exhaling, make a humming sound. Repeat this as many times as possible for the best results.

What Does Science Say About The Impact of Anger on Human Health?

Most of us have grown up listening to phrases linking an angry mindset to elevated blood pressure – **but are all these just old wives' tales, or is there a grain of truth in these remarks?** Several researchers have performed clinical studies that have in fact established a connection between anger and systemic disease.

For instance, Golden et al., in 2006, found through their study that higher levels of anger scores actually increased the chances of developing Diabetes by 34%! In addition, anger was also postulated to directly affect the cardiovascular system, leading to the release of excessive stress hormones like noradrenaline. The release of these factors may cause a sudden surge in energy levels, which may consequently lead to drastic repercussions on the dynamics of blood flow and metabolism.

On a more subtle level, the manifestation of anger has also been reported to promote the adoption of various unhealthy lifestyle choices such as smoking, alcoholism and caffeine consumption, etc[1-3].

This even leads one to wonder, if the systemic disease is a direct bodily consequence, or it is through the adoption of these harmful lifestyle habits due to an overall unhealthy mind-set that it leads to these manifestations.

Golden et al., in 2006, found through their study that higher levels of anger scores actually increased the chances of developing Diabetes by 34%.

Is Spirituality a Form of Anger Management?

It couldn't possibly be a coincidence that the wise old sages depicted in epics and tales tend to be serene, peaceful beings, right? Practices such as yoga and meditation have long been considered to be of paramount significance in reducing stress levels and promoting peace of mind. And while the stories are age-old, the science to support them may have emerged only a few centuries ago – meditation and yogic practices do indeed enhance our abilities to maintain a calm mind-set.

Yogic practices and meditation enhance our abilities to maintain a calmer mentality

A study group led by L Narasimhan has extensively studied the effect of integrated yogic practices on negative emotions like anger. They have highlighted in their research, the importance of mindfulness and yoga in managing anger and curbing its effects.

They actually noted a 45% decrease in hostility after yogic practices. They also witnessed dramatic reductions to a similar extent in feelings such as being upset or distressed, which are often considered different facets of anger due to unsatisfied desires or an inability to cope[4].

A detailed description of the Integrated Approach to Yoga Therapy (IAYT) has been given by the research group led by Mani. Their study revolved around developing and validating a module based on yogic practices for anger management in adolescents. Sage Patanjali has stated that asana practice effortlessly calms the mind.

Similarly, Pranayamas help promote slow and deep breathing patterns and restore nervous system balance, thereby tranquilizing the mind. Patanjali Yoga Sutras (PYS) and Yoga Vasistha (YV) have provided several techniques for anger management. PYS recommends distraction (*pratipaksha bhavana*) and physical activity (*asanas*) along with regulated breathing (*pranayama*). Yoga Vasistha highlights the sublimation of thoughts (*mana prashamana*)[5].

In the year 2019, Hagen et al. collected qualitative data in their European research project based on experiences shared by youngsters after being exposed to yoga intervention. Maharishi Patanjali, a renowned codifier of Yoga, promoted the practice of Yoga Sutras and also incorporated pranayama, asana, and meditation in this interventional research project.

The individuals selected were young people in their late teens or early twenties with Norwegian and refugee backgrounds. These individuals reported improved sleep habits, enhanced ability to regulate and cope with stress, and benefits in terms of refugee trauma.

Pause and Breathe

An individual with Attention-deficit hyperactivity disorder (ADHD) and anger issues shared his feedback stating how focusing on only his breath when he felt extremely frustrated helped him overcome the feeling of anger.

The "Pause and Breathe" technique, especially when the mind seemed angered by something, did help many of the participants in this study[6].

Raj-yoga meditation is a simple behavioral intervention wherein an individual sits in a relaxed posture with eyes open and fixes their gaze on a meaningful object. Whenever the mind seems to wander away, it is brought back by the object in regard[7]. Many studies have found the positive effects of this form of meditation, not just on the mind-set but also in terms of clinical outcomes in heart disease[8-10].

How does Goal-Directed Positive Self-Talk Help?

Have you noticed the way you talk to yourself when you're angry, upset, distressed, or overcome by any negative emotion stemming from a feeling of rage? Spontaneous self-talk in situations eliciting anger usually tends to be negative or retrospective ("I messed up").

As we now know how important our conversations with our own selves are, imagine what an impact they may have during such instances of heightened emotion!

Positive self-talk, on the other hand, has been linked to better performance and well-being in similar situations. Encouraging results have been seen with the use of motivational and instructional cue words on performance and performance-related outcome measures, and this has been demonstrated particularly in the field of sports[11]. Think of how many movies have depicted a last-minute surge of optimism due to a handful of positive words from the coach or a well-meaning friend and how they transformed the outcome of the game!

It is important to indulge in the habit of positive self-talk to overcome the ebb and flow of negative phrases in our minds due to negative emotions such as anger. **How do we do this?** By cleverly rephrasing these phrases!

Anger, similar to anxiety, is an emotion that can impact performance negatively. Being a high-arousal emotion, it can be advantageous in gross tasks requiring effort, power, or endurance, but can be disadvantageous in fine tasks that demand accuracy and precision.

Incorporating the aforementioned breathing techniques

and calming self-talk into our lifestyles, can help bring more harmony not only to our minds and bodies but also to the execution of our personal and professional undertakings.

Negative self-talk stemming from distressing emotions such as anger can be damaging to our performance. It is important to replace these with statements infused with positivity!

For every minute you remain angry, you give up sixty seconds of peace of mind
~ Ralph Waldo Emerson

Would You Like to Read Further?

1. Golden SH, Williams JE, Ford DE. Anger temperament is modestly associated with the risk of type 2 diabetes mellitus, The atherosclerosis risk in communities study. Psychoneuroendocrinology. 2006;31(3):325–332.

2. L. Hendricks, D. Aslinia. The effects of anger on the brain and body. Natl. Forum J. Couns. Addict., 2 (2013), pp. 1-12

3. Taggart P, Boyett MR, Logantha S, Lambiase PD. Anger, emotion, and arrhythmias: From brain to heart. Front Physiol. 2011;2:67.

4. L. Narasimhan, R. Nagarathna, H. Nagendra. Effect of integrated yogic practices on positive and negative emotions in healthy adults. Int J Yoga, 4 (2011), pp. 13-19

5. Tl AM, Sn O, Sharma MK, Choukse A, Hr N. Development and validation of Yoga Module for

Anger Management in adolescents. Complement Ther Med. 2021;61:102772.

6. Hagen I, Skjelstad S, Nayar US. "I Just Find It Easier to Let Go of Anger": Reflections on the Ways in Which Yoga Influences How Young People Manage Their Emotions. Front Psychol. 2021 Nov 22; 12:729588.

7. Telles S, Desiraju T. Autonomic changes in Brahmakumaris Raja yoga meditation. Int J Psychophysiol. 1993; 15:147–52.

8. Jain P, Kiran U, Saxena N. Anaesthetic management of a child with long Qt-Interval syndrome. Indian J Anaesth. 2002; 46:395–7.

9. Gupta SK, Sawhney RC, Rai L, Chavan VD, Dani S, Arora RC, et al. Regression of coronary atherosclerosis through healthy lifestyle in coronary artery disease patients – Mount Abu Open Heart Trial. Indian Heart J. 2011; 63:461–9.

10. Manchanda SC, Narang R, Reddy KS, Sachdeva U, Prabhakaran D, Dharmanand S, et al. Retardation of coronary atherosclerosis with yoga lifestyle intervention. J Assoc Physicians India. 2000; 48:687–94.

11. Hatzigeorgiadis A, Zourbanos N, Galanis E, Theodorakis Y. Self-Talk and Sports Performance: A Meta-Analysis. Perspect Psychol Sci. 2011;6(4):348-356.)

5

Fear

"*Our deepest fear is not that we are inadequate. Our deepest fear is that we are powerful beyond measure.*"

— Marianne Williamson

"What We Think, We Become"

What is this emotion, fear? Fear is a natural human emotion. It is not wrong to claim that it is an alerting system that alerts us of danger or threats which might occur physically or psychologically.

Have you ever pondered on a negative thought of fear for many days, which might have started with a conversation with another person? You might retire to your bed earlier in such cases as the fear might have exhausted you and might have drained you of your energy. We might not realize the consequences of fear as it always performs at the subconscious level. When it enters us, we enter into a realm of illusion where we create a bubble filled with fearful thought forms. Hence, at last, this illusion becomes a reality. When we try to break out of this bubble, we tend to believe that the life inside the bubble drenched with fear is the true reality and completely break away from the actual reality.

Pause for a second and ask yourself the following questions. **Are you living inside this illusionary bubble that you have created unconsciously? Are you making this fear factor stronger by trying to break the illusionary bubble? Are you unable to destroy this bubble with your energies?** We are unaware that we possess unbelievable power that, if appropriately sequenced, will give us unlimited strength to cope with the issues of fear with a positive mindset.

Note down the impact you have created for yourself through the energies you have made on yourself through the energies you have sequenced. Now, analyze the different life experiences you have gone through that aroused fear within you. Observe how you dealt with the fear experienced in each situation. Has that fearful experience frozen your actual reality? If it has, then what are the processes involved in bringing you back from the illusionary reality to the original one?

"It is the unknown we fear when we look upon death and darkness, nothing more."

— J K Rowling

Step 1 AWARENESS

The first step involved is "Awareness". It is the foundation that enables us to fight against fear. Deprived of this, the individual will lose the power to identify the fear factor approaching their consciousness. "Stillness with deep awareness" is the primary step involved in the process of fear management.

This has to be practiced every day, either before retiring to bed or early morning, before starting any daily task.

Step 2 PROCESSING

Once the emotion of fear is recognised, process it by reasoning out how or what triggered this condition. It is a crucial step as it helps us gain wisdom over our actions and prevents us from repeating the actions that trigger fear. This is the phase where our awareness processes the different emotions involved through mindful meditation.

Step 3 BREATHING TECHNIQUE

Once clarity is gained on how and what triggered the factors that bind you with fear, you should disintegrate them to move it out of your emotional body. Taking deep breaths would help you in gaining clarity and get relief from the emotions of fear. For that, sit in a comfortable position either on a chair or on a mat with your legs crossed. Sit in an upright position, with your neck relaxed and eyes closed. Then, inhale and exhale deeply. While taking deep breaths, use the power of your awareness to identify the factors that create fear within you. The common, deluded, and unhealthy fears within us include fear of dying, fear of loss, fear of failure, etc. With deep consciousness, process the fear component to check the types of experiences that trigger fear.

Ultimately, through these analytical techniques, you will realize that your deluded minds and negative actions are the culprits behind your fear. You must observe the process and contemplate the gathered information to gain more wisdom on the emotion of fear. Visualize your fears clubbed with the actual causes of them in the form of dense and thick smoke. Breathe them out through the nostrils and mouth to help you get over your fear. According to the emotional basis of this technique, the individual throws away the factors of fear, which disintegrates and fades away, never to return. This paves the way for inhaling pure air, which symbolizes inspiring, calm energy that stands for fearlessness in the form of blissful white light that engorges our body and mind.

It is mandatory to repeat this process of inhalation and exhalation until your mind becomes light and subtle. A clear, peaceful, and fearless mind is the foundation of a healthy body and healthy emotional stability.

Neurological Basis of Fear

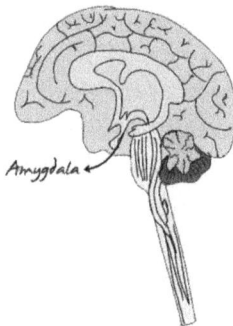

The amygdala plays a key role in regulating fear and anxiety.

Fear in the social setting manifests in a variety of different ways – anxiety, phobias, and even full-fledged panic attacks – and can have a tremendous negative impact on our personal and professional undertakings. It can be described as an automatic neurophysiological state of alarm, which is characterized by a 'fight-or-flight' response to a cognitive appraisal of present or imminent danger that may be real or perceived[1]. **What does that imply?** That one may produce the physical and psychological manifestations of the emotion of fear, even in the absence of any real cause of alarm! When there is an overestimation of this perceived 'threat' or an erroneous judgment of imminent danger, it can cause an inappropriate or exaggerated response, resulting in a pathological state of anxiety.[1]

The amygdala plays a key role in regulating fear and anxiety. Individuals with anxiety disorder tend to display heightened amygdala responses to perceived threats. Pharmacological interventions act by suppressing the connections of the amygdala and the limbic system to the prefrontal cortex and thus provide symptomatic relief by attenuating this response.[1] While it may be a mental health disorder, mortality rates are higher in patients with anxiety due to the risk of suicide and even adverse cardiac events.[1,2] This gives the phrase 'being scared to death' a whole other meaning.

Social Impact of Fear and Anxiety

"Of all the liars in the world, sometimes the worst are our own fears."

— Rudyard Kipling

Fear – while beneficial in moderate amounts under certain circumstances – can severely impact our performance in situations that require a great deal of precision and accuracy. Anxiety has been noted to severely impact professional performance – moreover, the consumption of medication used to treat it is also associated with impairment of activity at work. A proportion of patients were also found to be unaware of the initial worsening of symptoms associated with the intake of medication.[3] Moreover, owing to the stigma related to mental health problems, employees find it difficult to disclose their diagnoses to colleagues and supervisors, or even family and friends, thus negatively impacting interpersonal relationships as well.

Physiological Impact of Mindfulness Techniques to Overcome Fear

While pharmacological interventions are mandated in an acute setting, there has been some evidence to support the incorporation of meditation techniques in the treatment of anxiety and other fear-based conditions. A team led by Zeidan employed 15 healthy subjects for four days of mindfulness meditation and found a beneficial impact on levels of anxiety after the practice of meditation. They also noted through their study that anxiety relief after meditation practice was associated with the activation of the anterior cingulate cortex, ventromedial prefrontal cortex, and anterior insula. They, therefore, concluded that mindfulness meditation stabilizes the mind by focusing attention, assessing certain events as temporary and curbing extreme reactions to situations.[4] Regular practice of meditation can thereby prevent abnormally heightened responses to situations that do not warrant it and enable us to make better, more mindful decisions.

Another team led by Koyuncu and Bulbul conducted a study to assess the effect of yoga on fear of childbirth – undoubtedly something that stems from our fear of bodily pain in general. Their methodology consisted of three stages – Prayanama (breathing techniques), Asanas (poses and body movements), and Meditation (focus). They randomized pregnant women into two groups – the first group was subjected to yoga practices, whereas the second group did not follow any. They found that the practice of yoga significantly reduced fear, anxiety, and other negative emotions in pregnant women in the first group when compared to the women in the second group. Childbirth self-efficacy was also found to be higher in the first group than in the second.[5] Could it be that regular practice of meditation and a combination of powerful asanas helped these women focus on the joy they were bringing into this world and overcome their fear of the temporary process heralding it?

The practice of yoga significantly reduces fear, anxiety, and other negative emotions in pregnant women.

A randomized control trial conducted by a group led by Elizabeth studied the effect of mindful meditation on anxiety. They performed a study wherein individuals were randomized to an 8-week group intervention or to an attention-control stress management education. A validated measure of anxiety levels, the Hamilton anxiety scale, was used to measure symptoms associated with anxiety. They actually found that meditation techniques significantly curbed symptoms of anxiety and led to lower distress levels, along with improved reaction to stress and coping[6]. So you see, in addition to reducing levels of negative emotions, these practices also promote better and healthier mechanisms of dealing with them, should they arise.

The positive effects of mindfulness-based interventions (MBI) were also highlighted by a group of researchers led by Liu. They conducted a study among subjects who were diagnosed with a social anxiety disorder. This is a condition where emotions such as fear and anxiety are so dominant they disrupt one's relationships, daily routines, work, school, and other regular activities. These researchers studied the effect of mindfulness-based interventions amongst these subjects and concluded that they were superior to the no-treatment group. They also discovered that the effects of these interventions lasted for about 12 months![7]

Mindfulness-based techniques and meditation practices have been documented to have a positive impact on the treatment of symptoms related to anxiety across ages, genders, races, and populations. As we discussed earlier, the utilization of pharmacotherapy in situations that warrant it – especially acute, emergent, and life-threatening conditions – is absolutely essential and its role remains undisputed. However, considering the chronic nature of this condition and the large impact of its manifestations on a person's psyche, the incorporation of these practices in conjunction with pharmacotherapy as required, could potentially translate to better patient outcomes and improved standards of living.

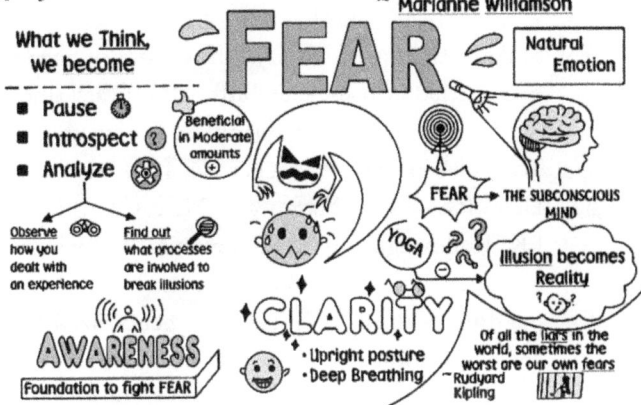

Our deepest fear is not that we are inadequate.
Our deepest fear is that we are powerful beyond measure.
~ Marianne Williamson

What we Think, we become

FEAR

Natural Emotion

- Pause
- Introspect
- Analyze

Beneficial in Moderate amounts

Observe how you dealt with an experience

Find out what processes are involved to break illusions

FEAR → THE SUBCONSCIOUS MIND

YOGA

Illusion becomes Reality

AWARENESS
Foundation to fight FEAR

- Upright posture
- Deep Breathing

Of all the liars in the world, sometimes the worst are our own fears
~ Rudyard Kipling

Would You Like to Read Further?

1. Chand SP, Marwaha R. Anxiety. [Updated 2022 May 8]. In: StatPearls [Internet]. Treasure Island (FL): StatPearls Publishing; 2022 Jan-. Available from:
    https://www.ncbi.nlm.nih.gov/books/NBK4703 6/

2. 2. Pereira AS, Willhelm AR, Koller SH, Almeida RMM. Risk and protective factors for suicide attempt in emerging   adulthood. Cien Saude Colet. 2018 Nov;23(11):3767-3777.

3. Haslam C, Atkinson S, Brown SS, Haslam RA. Anxiety and depression in the workplace: effects on the individual and organization (a focus group investigation). J Affect Disord. 2005;88(2):209-215.

4. Zeidan F, Martucci KT, Kraft RA, McHaffie JG, Coghill RC. Neural correlates of mindfulness meditation-related anxiety relief. Soc Cogn Affect

Neurosci. 2014 Jun;9(6):751-9. doi: 10.1093/scan/nst041.

5. S. B., & Bülbül, M. (2021). The impact of yoga on fear of childbirth and childbirth self-efficacy among third trimester pregnants. Complementary Therapies in Clinical Practice, 44, 101438.

6. Hoge EA, Bui E, Marques L, Metcalf CA, Morris LK, Robinaugh DJ, Worthington JJ, Pollack MH, Simon NM. Randomized controlled trial of mindfulness meditation for generalized anxiety disorder: effects on anxiety and stress reactivity. J Clin Psychiatry. 2013 Aug;74(8):786-92.

7. Liu X, Yi P, Ma L, Liu W, Deng W, Yang X et al. Mindfulness- based interventions for social anxiety disorder: A systematic review and meta-analysis. Psychiatry Research. 2021.Vol 300.

6

GRATITUDE

I am thankful for....

"Reflect upon your present blessings, of which every man has plenty; not on your past misfortunes, of which all men have some."
— Charles Dickens

Thank you - these two words contribute to one of the highest forms of positive vibrations. In fact, this phrase of

Gratitude uplifts us energetically to perceive life in a positive way.

First and foremost, we should acknowledge the fact that all the experiences we experience in our life are the outcomes of our own doing. The phrase, "It is our doing," moves us into a dimension of stress and tension along with anxiety.

However, always remember that we, human beings, see the smaller picture. When things don't go our way, it results in low self-esteem and fear of failure. Hence, it is very important to train ourselves to connect to the DIVINE. It is a truth worth upholding that a powerful force in the universe keeps us all in balance. Gaining ardent faith in this aspect is crucial, and hence, make it a point to stress the connection between yourself and the DIVINE in your life.

Even though our vision is limited, the universal energy force sees a larger picture unseen or unperceivable. Building faith in this dimension of existence helps us to surrender to this divinity. The truth is, "Energies work like magic when you have complete faith in it."

- STEP 1. Build faith in the divine existence
- STEP 2. Learn to surrender to these energies.
- STEP 3. GRATITUDE: Gratitude will flow naturally once we have built faith and learned how to surrender to this Divine Force.

When these three steps of FAITH, SURRENDER, and GRATITUDE are in sequence, move with them throughout your life. In fact, we gain unbelievable strength and power to withstand any challenging situation that comes our way.

Cultivating the power of Gratitude in our journey of life allows us to release ourselves from the bondage of future moments that is in the control of a higher power. That is why it is said that when good times come, say "thank you" and move ahead; likewise, when bad times come, say "thank you" and grow.

Every challenging situation helps us to evolve with wisdom and strength. These are the treasures we need to recognize in our journey and use in our future experiences. Hence, always make a point to say thank you and keep moving ahead.

Thank you - these two words need to be repeated again and again like a mantra in your mind; so that it starts becoming a subconscious ritual in your mind. Once you have made this mantra a subconscious act, then this vibration flows into every activity in your life. The result of our actions will be filled with this powerful energy. In this way, we are never worried about the fruit of our activities as the fruit comes very naturally to us.

My father once told me how he used Gratitude in his life. He said his workplace has so many stairs and it was necessary for him to walk up and down the stairs for his hospital rounds (He chose to take the stairs rather than the elevator). He mentioned that while taking every step, he said "thank you" in his mind. This was an eye-opener for me toward the energy of Gratitude. We need to work on this to such an extent that the vibration flows with our inhalation and exhalation.

- I am grateful for my good health.
- Thank you for my balanced mind and emotions
- I am grateful for my loving family and friends and everything they do for me.

- Everything in life is working 'for' me…Thank You.

How to Perform Gratitude Meditation?

Tip: Keeping a gratitude journal helps you focus on your blessings – and also gives you something to turn back to on a rainy day when you're too low to find something to be grateful about!

Maintain a gratitude journal and work on it daily. Add a few instances from your daily life which inspired a "thank you" feeling within you. In fact, affirmations, along with Gratitude, work tremendously on our energy levels. It is mandatory to practice affirmation techniques to buck up your energies. Try saying: I am strong, I am confident, I am happy, I am amazing, I am great, I am healthy, I am unique, I am special, I am gifted, I am loved, I am loveable, I am joyous, I am fabulous, I am wonderful and above all… I am ME.

Our words have the power to transform our reality, as it aids in changing the way we think, feels, and vibrate. There have been several studies conducted on the impact of pictures, phrases, and music on the state of frozen water. In fact, frozen water takes different forms mainly depending on whether it was exposed to positivity or negativity during the freezing process. This study represented a great step that offered a scientific basis for establishing the spiritual beliefs that were in vogue earlier. If the study is true in the case of water, just imagine how much impact your words will have on other people, as well as on your own thoughts, feelings, and potential, as the human body is made up of 70%-72% of water.

Can We Define What It Means to be "Grateful"?

Things I am grateful for

My Parents
My Pet
Music
Sunsets
My home
Rasberries
Chocolates
Gardening
Walking with my grandpa
Art
Sea
Harry Potter
Scrabble
Nutella
My school
Paints

"*I cursed the fact I had no shoes until I saw the man who had no feet.*"

— Persian proverb

Gratitude has been defined as an attribution-dependent state that is a result of a two-step cognitive process – the first step being recognition of a positive outcome and the second one being recognition of the fact that there is an external source for this positive outcome.[1]

It has been a core component of most religions, such as Hinduism, Islam, Buddhism, and Christianity, and the trait has been very highly regarded across many cultures. It is no wonder, then, that several cultures have holidays dedicated to the expression of Gratitude, and it is one of the earliest and most important values taught to children by their caregivers.

However, research suggests that Gratitude is not merely

a social or cultural construct. In fact, it has deep roots that are entrenched in our evolutionary history – not simply in our sociocultural and biological development, but also in our DNA. Whether it is an emotion, a trait, a mood, or a state of being, it is found to be pervasive across various species – fish, birds, and even vampire bats have been noted to engage in "reciprocal altruism," which is viewed as an offshoot of Gratitude.

What Do We Mean by "Reciprocal Altruism"?

> "The only people with whom you should try to get
> even are those who have helped you."
> — John E. Southard

What exactly is "reciprocal altruism"? The term encompasses all activities or behaviors that an animal may perform to help another member of its species, at times at a cost to their own well-being, seemingly because they appear to identify on a visceral level that the individual may repay the favor in the future. This desire to repay kindness is viewed by some scientists as an expression of Gratitude – some even suggest that Gratitude may have evolved as a mechanism to drive reciprocal altruism and forge stronger emotional connections between individuals.[2]

The vampire bat is among several species of animals that demonstrate 'reciprocal altruism', a trait linked to an attitude of gratitude

The "Neurobiology" Behind Our Altruism

Amygdala is the neurological center for responding to real or perceived 'threats' and is considered the 'neural alarm' system.

It plays a critical role in up-regulating threat-related responses – such as 'inflammation,' the body's response to injury or infection. Cultivating a state of Gratitude has been noted to produce a reduction of the stimulated release of the inflammatory marker TNF-α by 'down-regulating' or reducing the activity of the amygdala. This has been shown to attenuate systemic inflammation that could lead to different forms of illness or chronic disease.[3-5]

There have been other studies as well, which have implied a plausible role of inhibiting the amygdala in this grateful or 'giving' behavior. Some animal studies have demonstrated that the neural region known as the ventral striatum (VS) is responsible for the maternal 'caregiving' behavior, whereas the part of the brain known as the septal area (SA) helps in the reduction of emotions such as fear by inhibiting the amygdala.

The activity of both these regions was noted to be increased with emotions like Gratitude or caregiving to loved ones. In fact, researchers even noted that maternal instincts could be adversely impacted by lesions of the VS and SA.[6,7]

Besides these, Gratitude also results in the activation of multiple other neural regions, like the brain's reward pathways and the hypothalamus.

It can augment the release of the neurotransmitter serotonin, which in turn stimulates the brainstem to produce dopamine, the brain's pleasure neurotransmitter associated with feelings of well-being.[8]

The research group led by Tani, observed that Gratitude has a positive effect on the cognitive function of human brains — or in other words, the ability to learn, think, memorize, reason, solve problems, focus, and make decisions. This was evidenced by the presence of large volumes of the amygdala and fusiform gyrus on the MRI (Magnetic Resonance Imaging) associated with higher levels of Gratitude.[9]

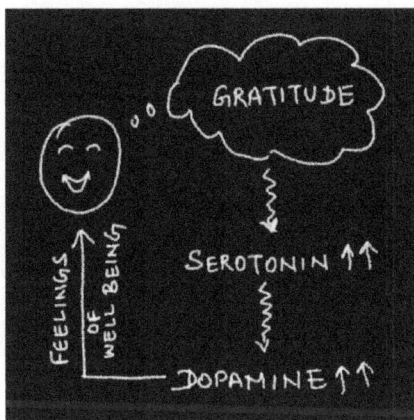

Gratitude promotes the release of the brain's 'pleasure neurotransmitter' dopamine – leading to a positive impact on mental and emotional well-being!

Cultivating Gratitude for Healthful Living

Research states that Gratitude is not only crucial for neurological health but is also associated with the maintenance of optimal systemic health. **But does physiological well-being simply refer to good physical**

health? It embodies so much more – it includes a positive impact on subjective and objective well-being, an optimistic outlook on life stemming from a healthy state of mind, and wholesome social relationships.

Researchers have demonstrated the role of Gratitude in producing a positive impact on all of these factors by virtue of mu opioid signalling.[10] They performed a systematic review of the effects of 'gratitude interventions' – or activities that promoted grateful behavior – on parameters of physical health. Remarkably, they found improvements in several of them – lowering of blood pressure, control of asthma, regulation of blood sugar levels, and development of healthy eating habits.[11]

A team of researchers led by Piferi suggested that individuals who possess the trait of 'giving to others' or Gratitude have comparatively lower levels of blood pressure than those who exhibit lower levels of this quality. In other words, the state of Gratitude indirectly influences the cardiovascular well-being of individuals.[12]

What's more, nurturing a grateful attitude was also shown to promote good quality of sleep by preventing a negative pre-sleep thought process and facilitating positive thoughts before falling asleep.[13]

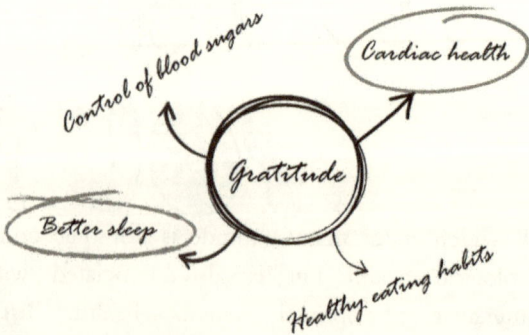

Control of blood sugars · *Cardiac health* · *Gratitude* · *Better sleep* · *Healthy eating habits*

Gratitude interventions have been linked not only to
more positive mental health but beneficial systemic
outcomes too!

A research experiment led by Warber examined the
effects of a 'spiritual retreat' on depression and other such
measures of well-being in post-acute coronary syndrome
patients. Gratitude interventions, such as guided imagery,
journal writing, meditation, and nature-based activities, were
incorporated into a four-day spiritual retreat, and the
patients assigned to this group showed significantly lower
depression scores that were retained for up to three months
as compared to the group that was not included in the
retreat.[14]

There is extensive scientific evidence to back up the
positive impact that promoting a grateful attitude has, not
merely on our mental well-being but also on our physical
and physiological health, translating to lower cellular
inflammatory index, better cardiovascular health, optimal
sleep patterns, etc. It is more than just a fleeting enjoyable
emotion – it bears the potential to transform an individual's
existence. It may well be considered one of the most
important values to be developed and cherished to promote
health and happiness among all.

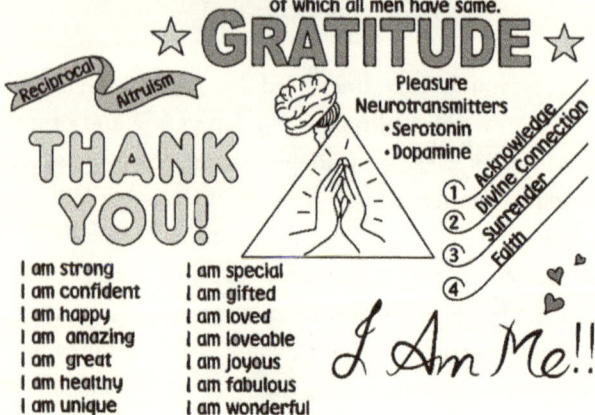

Would You Like to Read Further?

1. Emmons RA, McCullough ME. Counting blessings versus burdens: an experimental investigation of gratitude and subjective well-being in daily life. J Pers Soc Psychol. 2003;84(2):377-389.

2. Trivers R. L. 1971. The Evolution of Reciprocal Altruism. Quarterly Review of Biology 46: 35–57

3. Adolphs R, Tranel D, Damasio H, Damasio AR. Fear and the human amygdala. J Neurosci. 1995;15(9):5879-5891.

4. Eisenberger NI, Cole SW. Social neuroscience and health: neurophysiological mechanisms linking social ties with physical health. Nat Neurosci. 2012;15(5):669-674.

5. Hazlett LI, Moieni M, Irwin MR, et al. Exploring neural mechanisms of the health benefits of gratitude in women: A randomized controlled trial.

Brain Behav Immun.   2021; 95:444-453.

6.   Hansen S. Maternal behavior of female rats with 6-OHDA lesions in the ventral striatum: characterization of the pup   retrieval deficit. Physiol Behav. 1994;55(4):615-620.

7.   Strathearn L, Fonagy P, Amico J, Montague PR. Adult attachment predicts maternal brain and oxytocin response   to   infant   cues. Neuropsychopharmacology.   2009;34(13):2655-2666.

8.   Gratitude and the Brain: What is Happening? [Internet]. [cited 2022 May 14]. Available from: https://www.brainbalancecenters.com/blog/gratit ude-   and-the-brain-what- is-happening

9.   Tani Y, Koyama Y, Doi S, et al. Association between gratitude, the brain and cognitive function in older adults: Results   from the NEIGE study. Arch Gerontol Geriatr.   2022; 100:104645.

10.   Henning M, Fox GR, Kaplan J, Damasio H, Damasio A. A Potential Role for mu-Opioids in Mediating the Positive   Effects   of   Gratitude. Front Psychol. 2017; 8:868.

11.   Boggiss AL, Consedine NS, Brenton-Peters JM, Hofman PL, Serlachius AS. A systematic review of gratitude   interventions: Effects on physical health and health   behaviors. J Psychosom Res. 2020; 135:110165.

12.   Piferi RL, Lawler KA. Social support and ambulatory blood pressure: an examination of both receiving and giving. Int J Psychophysiol. 2006; 62(2):328-336.

13.   Wood AM, Joseph S, Lloyd J, Atkins S. Gratitude

influences sleep through the mechanism of pre-sleep cognitions. J Psychosom Res. 2009; 66(1):43-48.

14. Warber SL, Ingerman S, Moura VL, et al. Healing the heart: a randomized pilot study of a spiritual retreat for depression in acute coronary syndrome patients. Explore (NY). 2011;7(4):222-233.

7

VIBRATION

"How you vibrate is what the universe echoes back to you in every moment."

— Panache Desai

Your mind is one of the most powerful tools you can possess. It can either be a faithful servant or a dangerous master. It is crucial to choose wisely because the Universe responds to your frequency. It does not recognize your personal desires, wants, or needs.

It only understands the frequency at which you are vibrating. For example, if you are vibrating in the frequency of fear, guilt, or shame, you will attract things of similar vibration. If you are vibrating in the frequency of love, joy, and abundance, you will attract things that support that frequency.

It is like tuning into a radio station. You have to be tuned to the music you want to listen to, just like you have to be tuned to the energy you want to manifest into your life. Change your mindset, and it will change your life.

Reflection

> *"Focus on an ocean of positives, not a puddle of negatives."*
> — Kevin Ansbro

The most crucial step is to notice and reflect on what frequency you are in, to vibrate higher. Peter Drucker, the father of modern management, once said: "Follow effective action with quiet reflection. From the quiet reflection will come even more effective action". Clearly, the master realized the deep connection between taking time for reflection and vibrating higher.

All the actions you undertake in life begin and end in your mind. The emotions and feelings to which you give power, in turn, have power over you, whereas the ones you left back will never show dominance over you. You might also wonder about the different types of attitudes you experience in people you meet. The attitude that they show to you is based on how they are treated by you and how they are treated by others. Hence, when it comes to attitude, always remember, "Don't blame the paint when you hold the brush."

Now, thinking about ourselves, we should remember that our ideas, our rearing, and our environment will project into our behavior. There is a saying, "What we do flows from what we are". Most of the time, people concentrate more on living conditions rather than on the real purpose and essence of life. It is a collective disease because we try to cure the symptom-living conditions rather than looking for the root cause of the disease, which is the essence of life.

Make sure to avoid taking in things and people whom you dislike. Such actions will only strengthen your power over yourself. Ikkyu Kensho Tzu once said, "When 'i' is talking, 'I' am listening. This is how I know who is talking." Hence, when it comes to sensitive issues, emotions, and feelings, always make sure that your inner 'i' is talking and the external 'I' is listening. Indeed, you should be thankful for the ones who triggered and tested you. They are the people that set you free from the patterned notions of life.

Do you think that happiness and joy rest only with a few people? If so, then you are wrong. It is not stagnant or pinned down to certain people and things. Happiness is portable; hence make it a point to carry it wherever you go.

I pledge myself to the purgation of all that is poisonous or illusionary within myself, and I propose to myself in the facility of truth in the world. I call out the cleansing and purification of all levels of my energy field and physical body, making it the perfect vehicle for love and light.

Powerful Quotes to Raise Your Vibrations

Let us look at some of the powerful quotes we can reflect on and imbibe in life.

There is nothing more significant to true development than understanding that you are not the voice of the mind - you are the one who hears it.

Eckhart Tolle once said, "*When you are detached, you gain a higher vantage point from which to view the events of your life instead of being trapped inside them.*"

Be open so that life will be easier. It is just like adding a spoon of salt to a glass of water in a lake. A spoonful of salt in a glass of water makes the water undrinkable. A spoon of salt in a lake is almost unnoticed.

"You are experiencing the echo of the vibration you are emitting."
— Creig Crippen.

Do you know the two most powerful words in the world? They are 'I AM.' They are the two most powerful words in existence because you become whatever words you put after them.

"If you want to know the secrets of the universe, think in terms of energy, frequency, and vibration."
— Nikola Tesla.

Life is not happening to you; life is 'responding' to you.

"Quantum physics tells us that nothing that is observed is unaffected by the observer. That statement, from science, holds enormous and powerful insights. It means that everyone sees a different truth because everyone is creating what they see."
— Neale Donald Walsch.

"How you do what you do is more important than what you do. The "how" refers to the underlying state of consciousness."
— Eckhart Tolle

The highest form of intelligence is the ability to observe without making any judgment. However, acquiring this intelligence isn't that easy. To do so, you will have to be in sync with your surroundings. In other words, you have to observe more and absorb less. Be a mere spectator and notice everything happening around you; just don't do anything about it.

Let's look at a few practices that help us be in sync and raise our positive vibrations.

- Loving
- Smiling
- Blessing
- Thanking
- Playing
- Painting
- Singing
- Dancing
- Meditating
- Yoga
- Tai Chi
- Daily workouts
- Being amidst nature
- Farming
- Consuming organic food

Positivity

Due to your ego-self, you often end up finding yourself in the victim's spot and its accompanying mentality. You become of the opinion that nothing in the world is sufficient to satisfy your needs. Your ego-self makes you believe that your existence is temporary, and this scares you. This impermanence also makes you believe that everything around you is complicated.

However, if you distance yourself from this negativity and condition your brain to think only positive thoughts, you will realize that you are the creator of your reality. Although you still have a fear of scarcity and living a temporary life, a positive mindset will make you enjoy everything while it lasts.

How do you develop this positive mindset?

Simple. Look for the elements that disturb your mind. These may be anything; people you want to stay away from, emotions and fears that are yet to be confronted, self-harming beliefs and attitudes, your fears, perspectives... anything.

Now, let's look at the roles of the conscious and subconscious mind in our daily lives. While the conscious mind needs to rest every day, our subconscious mind can stay awake all the time. Our subconscious mind helps us mull over various situations that were too much to handle for our conscious mind.

The subconscious mind allows you to walk down the path of earlier experienced spaces. This revisiting helps you in unveiling the truths that are different from your earlier revelations.

Now, let me ask you something. Whenever you think of something unfortunate, or an unfortunate incident happening to a loved one, you quickly divert your mind and do all sorts of weird things to prevent the Universe from listening to it. Or, let's say you acknowledge a good thing that has happened to you, and you quickly say, "touch wood" to avoid bad fate. I'm sure you have done this. And you are not wrong.

It is true that the Universe responds to your frequency. As they say, if you have positive thoughts, you attract positivity. If you have happy people around you, you will also be happy. However, if you think of negative thoughts all day long and surround yourself with gloomy people, you end up attracting negativity. So, the Universe responds more to your vibe than your desires and wants.

Don't believe me? Well, then do a quick test. Switch on the FM. Surf through the different channels. Did you come across a song you like? What song is it? Why do you like it? What is the mood of that song?

Your selection of the song solely depends on your thoughts. If it is a happy song, you have a positive attitude. However, if you choose a sad song, then you have a negative attitude. So, make a decision now. Do you want to stay positive and attract positivity or be negative and attract sad things?

> *"If you are positive, you will see opportunities instead of obstacles."*
>
> — Confucius

The Three Approaches

> *"The secret of change is to focus all your energy; not on fighting the old but on building the new."*
>
> — Socrates

Now that we have learned about positivity and negativity let's look at three major ways of approaching life:

- Passive approach
- Proactive approach
- Flow approach.

The passive approach is when you take life head-on. Just like in the film, 'Yes Man', say 'yes' to whatever comes your way and face it with lots of positivity. Once you get into the mindset of "Oh, this isn't going to work out," you will fail before even you start.

Most people give up soon in life, believing it's just their fate. Well, be assured that it is not. Navigate your life with a positive attitude, and your life will be smooth sailing.

All that said, the real need here is to identify why people tilt towards negativity whenever life throws hurdles at them.

Why do people resort to negative thoughts and not positive ones?

There are a couple of explanations for this.

Firstly, the people you are surrounded by matter. If your family and friends are living their life whining and complaining all the time, you will also become one of them without your knowledge. Secondly, you might be reacting a particular way to the situation because of your past experiences. Your past behaviors might come into play and rule your present actions. Therefore, try to surround yourself with happy thoughts and people and rule out negativity from your life.

The proactive approach is when you are taking full responsibility for your life, and you believe you are the co-creator of your life. You are willing to look deeply at what has gone wrong and are ready to make those changes in your life. You make a continuous effort to understand why things are happening the way they are happening in your life. You break the patterns and consciously create your reality.

The flow approach is a very mature way of dealing with life. Here, you are in complete harmony and in sync with everything. You are in tune with nature and are living in the moment. When you are living in the moment, only the present actions matter; your past actions will have no control over your life. This approach may sound like a passive approach, but it is very different. Here, you are completely in conversation with the Universe. You know about the consequences of your actions. Your life flows without any friction, and you are fully aware of the continuous dynamic exchange between you and the Universe.

How Does Music, Dance, and Meditation Affect Health?

A scoping review of 63 papers studied by multiple teams at the University of Brisbane, Australia, demonstrated that music and dance were beneficial in terms of health, recovery from illness, and overall development. Group singing supported cognitive health and well-being in adults with mental health illnesses, stroke, dementia, lung diseases, etc. They showed that children who learnt music or dance or movement in the form of yoga performed better in school and socially.[1] Finn and Fancourt et al studied the effect of music over stress-inducing/reducing biomarkers. They found that markers such as cortisol, blood glucose, and immune system measures changed in response to music demonstrating an overall reduction in stress levels.[2]

Dance for Parkinson's was a movement started in 2001 by Olie Westheimer. Parkinson's disease (PD) is a neuro-degenerative disorder that limits physical movements and cognition. A rigorous creative dance form was created for Olie's group members with PD and it helped them immensely. They reported physical, emotional, and social benefits. In a study done to assess the effects of dance in moderate to severe PD[3], it was realized that dances such as tango significantly improved the quality of life amongst these patients. Those bound to wheelchairs benefitted from group dance lessons than one-on-one lessons as it also allowed socialization.

It is one thing to be a patient admitted to a hospital, yet a whole other fiasco to be one who is mechanically ventilated. In intensive care units (ICU), patients are often restless, and we cannot blame them. They are connected to several lines; monitors beep all around them, they wouldn't know day from night and the absence of movement makes them feel worse. They are dependent on nurses and

caregivers for everything and those who are on mechanical ventilators are even robbed of their ability to speak. While all this is being done to save their lives and help them recover from deadly illnesses, the level of anxiety is bound to rise. Recently, several studies have been conducted to demonstrate the effect of music on such patients[4]. While it made no difference to their recovery period or mortality, it definitely helped reduce anxiety in them. Music soothed them and even reduced the number of sedatives required to calm them down. It is now considered a prescribed form of treatment for patients on ventilators who seem agitated and anxious.

We have heard and read a lot about how meditation has health benefits. It not only provides stress relief but allows faster and better recovery in many kinds of illnesses – both physical and mental. A recent study by Tanuj Dada et al.[5], unearthed the benefit of mindfulness meditation in a sight-threatening disease called Glaucoma. Glaucoma is known as the silent thief of sight as it is asymptomatic till an advanced stage. It is common in the elderly and in those with a positive family history. The only modifiable risk factor in glaucoma in order to prevent it from progressing is intraocular pressure (IOP). A randomized control trial was done in 100 patients with moderate to severe Glaucoma. One group underwent mindfulness meditation (MM) for 3 weeks along with anti-Glaucoma medications while the control group was only on medications. It was found that MM helped in reducing intraocular pressure by more than 23%, while it was only 5.82% among the controls.

A Picture Speaks a Thousand Words

We have heard of great artists like Pablo Picasso, Vincent Van Gogh, and Edward Munch. **But did you know that their art spoke much more than just brush strokes and expressions?** Their art spoke of their mental make- up. They suffered from bipolar disorder, and their art was testimony to it. Picasso's "blue period" consisted of art using blue and colder colours, much like how he felt – blue – depressive, speculated to have been painted during the period he lost his close friend to suicide[6]. While art can be a form of expression of mental health, it can also be used as a therapy to cure mental illness.

Art therapy is well established now, and books on mindful art or Mandala art therapy can easily be found on most social markets. **But is there a neuronal connection between art and health?** There definitely is, and the study of it is called neuro-aesthetics. Researchers show that the reward system in our brain gets activated while creating or viewing art. It releases happy hormones such as dopamine and oxytocin that trigger pleasure and positive emotions.[7] Clinical research has also indicated that art therapy has a meaningful impact on military veterans suffering from a post-traumatic stress disorder and even traumatic brain injury. Experiences through art and music help them channel their aggression and anger into more healthy forms of self-expression.

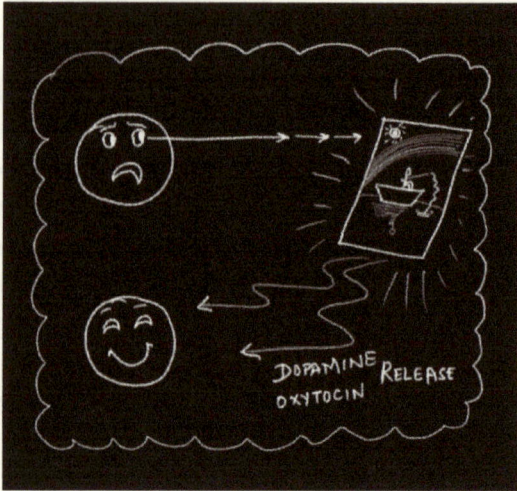

Does Science Relate with Positive Affirmations?

Self-affirmation theories focus on the premise that people are intrinsically motivated to maintain their self-integrity and when anything, for example, negative thoughts or judgment, goes against that, they feel threatened. Self-defense comes into play when one feels threatened. Self-affirmation exercises deal with reflection on meaningful values and bolster self-integrity and reduce self-defensiveness. The same was tested in a group of students at a university.[8] Participants who were made aware of self-affirmation and then told to self-affirm did not benefit as much as participants who were made aware but were given a free choice of the value they felt was meaningful. This goes to show that self-affirmation is something that can be made aware of in people but cannot be forced upon them as each person has different values integral to them.

In everyday life, we come across multiple stressors – from exams to job interviews to marriage prospects to board meetings or even parent-teacher meetings. At every stage, or in fact, even waking up every day with an "I AM" self-affirmation, can help in reducing the stress of the upcoming challenge or obstacle. Similar to stressors on the outside, there are stressors inside our body as well. Catecholamines or epinephrine and norepinephrine are our stressor hormones. They are responsible for the fight-or-flight response that we generate when we feel a threat. When in excess, they are excreted via urine.

The level of catecholamine in urine was studied by Sherman et al. during the period of exams in undergraduate students.[9] One group of students was given a self-affirmation exercise (essays) two weeks prior to the test, while the control group was not given any. 15-hour urine samples were collected two weeks prior and on the day of the exam from both groups of students and analyzed for the level of catecholamines.

As hypothesized, the level of epinephrine was higher during the exam than two weeks prior, but it was significantly higher in the control group. There was no difference found in the performance levels of the students during the exams, but those from group 1 reported feeling more confident during the exam than those in the control group. Self-affirmation not only boosts confidence but buffers stress responses, especially in psychologically vulnerable groups.

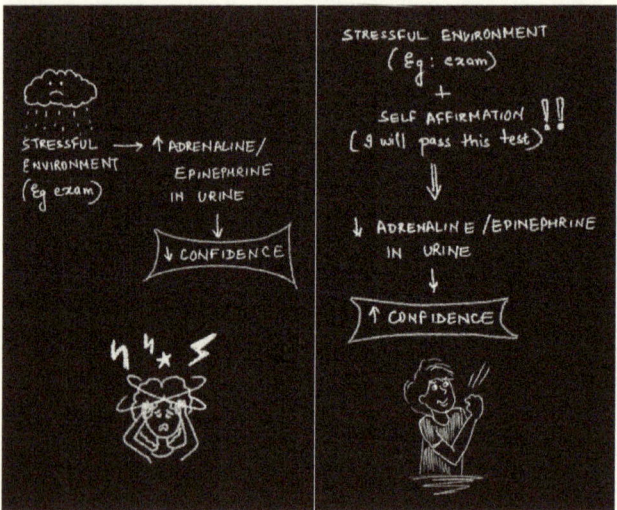

When we say positive attracts positive, we are not wrong. Christopher Cascio and his team examined the neural mechanisms of self-affirmation with a task developed for use in a functional magnetic resonance imaging environment. Participants who were self-affirmed showed increased activity in the prefrontal cortex as compared to those who were not affirmed, indicating that a positive affirmation has a direct impact on our conscious mind.[10] As shown by them, self-affirmations have benefits across threatening situations; affirmations can decrease stress, increase well-being, improve academic performance, and make people more open to behavioral change.[11]

On the contrary and true to our belief, persistent negative thinking or worrying can lead to negative outcomes. A systematic review of perseverative negative thinking revealed that it had an adverse effect on people with long-term health conditions. It burdened their mental health further by causing depression, anxiety, and emotional stress.[12]

Which Approach Would Lead to a More Satisfied Life?

Mihaly Csikszentmihalyi was the person who coined the term 'flow' and has researched the psychological effects of the flow approach extensively. It is a state of being that can be achieved no matter what task is being done[13]. To be able to enjoy a task, a person should be able to "be" at the moment.

The past or the future should not matter at that moment. It could be as simple as having a piece of chocolate cake without worrying about the weight that was lost the previous week or the weight that would be gained the next week. It could be as complicated as performing a crucial step in surgery without thinking about how the earlier step was done or about what complication could occur if this step is done wrong. Scholars have designed a work-related flow scale that involves assessing absorption at work, work enjoyment, and intrinsic motivation.

There is another flow state scale (FSS) that assesses components of flow, such as the merging of action and awareness, a high sense of control of one's actions, loss of self-reflective consciousness, and clear feedback. What these scales mean to tell us is that the flow approach is positively related to job performance.

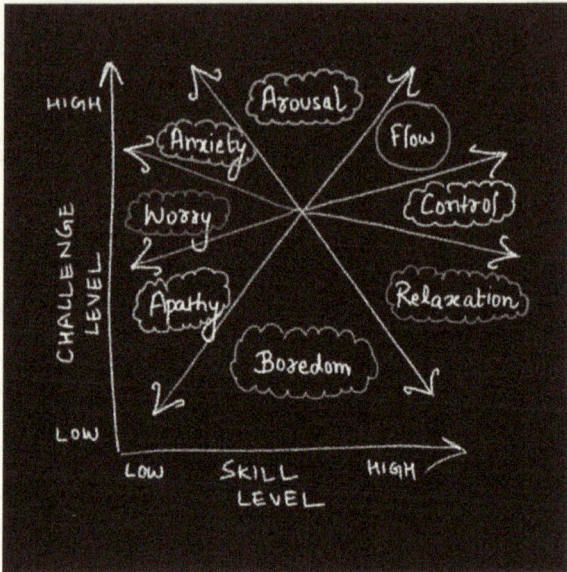

While the three approaches – passive, proactive, and flow are not totally independent of each other, each can benefit an individual in different ways. Some authors argue that to be successful, a proactive approach is important and that it is a predictor of work-related performance.

Arnold B Bakker proposed self-determination strategies to be proactive in the workplace.[14] These strategies include self-leadership, job crafting, playful work design, and the use of strengths. Bakker concluded that workers who are in flow experience greater pleasure and are intensely involved in their activities, but proactive self-determination could satisfy their psychological needs and contribute to their flow state of mind. The effect of proactive decision-making was studied by Siebert et al. in 2020. They established a strong correlation between proactive decision-making and general self-efficacy.

They implied that individuals could have a positive effect on their life satisfaction by deliberately choosing to follow a more effective decision-making approach. A goal-directed behavior driven by effective decision-making is a meaningful determinant of life satisfaction.[15]

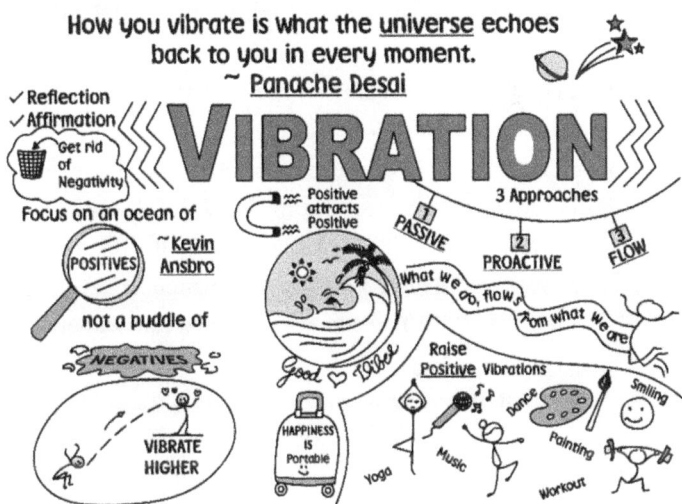

Would You Like to Read Further?

1. Finn S, Fancourt D. The biological impact of listening to music in clinical and nonclinical settings: A systematic review. Prog Brain Res. 2018;237:173-200.

2. Hackney ME, Earhart GM. Effects of dance on balance and gait in severe Parkinson disease: a case study. Disabil Rehabil. 2010;32(8):679-84

3. 4. Bradt J, Dileo C. Music interventions for mechanically ventilated patients. Cochrane Database Syst Rev. 2014;2014(12):CD006902. doi: 10.1002/14651858.CD006902

4. Dada T, Bhai N, Midha N, Shakrawal J, Kumar M, Chaurasia P, Gupta S, Angmo D, Yadav R, Dada R, Sihota R. Effect of Mindfulness Meditation on Intraocular Pressure and Trabecular Meshwork Gene Expression: A Randomized Controlled Trial. Am J Ophthalmol. 2021 Mar; 223:308-321.

5. Hajar R. Despair and Hope. Heart Views. 2017 Apr-Jun;18(2):66-67

6. Magsamen S. Your Brain on Art: The Case for Neuroaesthetics. Cerebrum. 2019 Jul 1;2019:cer-07-19. PMID: 32206171; PMCID: PMC7075503.

7. Silverman, Arielle & Logel, Christine & Cohen, Geoffrey. (2013). Self- affirmation as a deliberate coping strategy: The moderating role of choice. Journal of Experimental Social Psychology. 49. 93–98. 10.1016/j.jesp.2012.08.005.

8. Sherman DK, Bunyan DP, Creswell JD, Jaremka LM. Psychological vulnerability and stress: the effects of self-affirmation on sympathetic nervous system responses to naturalistic stressors. Health Psychol. 2009 Sep;28(5):554-62

9. . Cascio CN, O'Donnell MB, Tinney FJ, Lieberman MD, Taylor SE, Strecher VJ, Falk EB. Self-affirmation activates brain systems associated with self-related processing and reward and is reinforced by future orientation. Soc Cogn Affect Neurosci. 2016;11(4):621-9

10. Cohen G.L., Sherman D.K. (2014). The psychology of change: self- affirmation and social psychological intervention. Annual Review of Psychology, 65, 333–71

11. Trick L, Watkins E, Windeatt S, Dickens C. The association of perseverative negative thinking with depression, anxiety and emotional distress in people with long term conditions: a systematic review. J Psychom Res. 2016; 91:89–101

12. . Ottiger, B., Van Wegen, E., Keller, K. et al. Getting into a "Flow" state: a systematic review of flow experience in neurological diseases. J NeuroEngineering Rehabil 18, 65 (2021)

13. Bakker, A.B., van Woerkom, M. Flow at Work: a Self-Determination Perspective. Occup Health Sci 1, 47–65 (2017)

14. Effects of proactive decision making on life satisfaction Siebert J.U., Kunz R.E., Rolf P.(2020) European Journal of Operational Research, 280 (3),1171-1187

15. Dingle GA, Sharman LS, Bauer Z, et al. How Do Music Activities Affect Health and Well-Being? A Scoping Review of Studies Examining Psychosocial Mechanisms. Front Psychol. 2021; 12:713818.

8

SELF-HEALING

"If you just allow your body and mind to rest, the healing will come by itself."

— Thich Nhat Hanh

How important is it to clean our bodies on a daily basis? How would you feel when you have not taken a shower for a day or two?

We give so much importance to our external body, but what do we consciously do to clean our minds daily?

Can you imagine the stagnant energies that we have been hoarding within us for years?

Just like the way we consistently wash our physical bodies, we should also wash off all the built-in negative energies within us daily.

When we do so, we can experience a cleansed emotional body, which is the most important aspect of a balanced mind, body, and soul.

For this process, your body needs to be completely relaxed and calm, which will help you process your emotions clearly.

Methodology:-
Healing the body MINDFULLY – Body scan healing technique

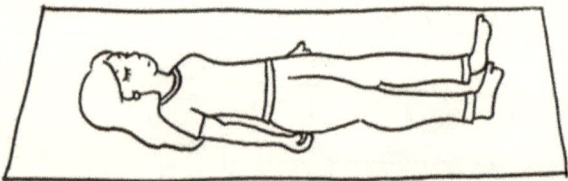

Every night just before you fall asleep, lie down on your back with palms slightly away from your body and facing upwards, feet slightly apart, and drop down your shoulders, close your eyes, and inhale and exhale deeply. First, bring your focus to your breath. Gradually, shift your awareness to your toes. Clench your toes and hold for a few seconds and then release them to relax every toe. The Affirmation to Relax should be said in your mind repeatedly. Now shift your awareness to the soles of your feet and relax them. Then move your attention to your ankles and completely relax your ankles. Affirm, "**My feet are healthy and relaxed**."

Shift your awareness to your calf muscles. Tense and squeeze your calf muscles and hold and affirm them to relax. Relax your shin and now tighten your knees, hold and relax. Tighten your thigh muscles next; hold, and relax. Affirm, "**My legs are completely healthy and relaxed**."

Then move towards your pelvic region- Affirm that all the organs within the pelvic region are healthy and balanced. Squeeze and tighten your pelvic muscles; hold and relax them. Now gently shift your mind towards your lower abdominal muscles and relax all the organs within this area. Pull your abdominal muscles in and hold for a few moments and then release and relax. Affirm, "**My abdominal region is healthy and relaxed**."

Now gently shift your awareness towards your chest region. Affirm, "**My chest muscles are relaxed, and all the organs within are healthy and functioning in a balanced manner**."

Now gradually shift your awareness towards your throat region and affirm, "**My neck muscles are relaxed, and the organs within are healthy and balanced**."

Now relax your hands, elbows, wrists, palms, and all your fingers. Make a tight fist; hold and then release and relax. Affirm, "**My hands are completely relaxed**."

Shift your awareness to your back. Affirm, "**My back muscles are completely relaxed**." Visualize your spine getting brighter and stronger. Then shift your awareness towards your head region. Tense your facial muscles; hold and release and relax. Maintain a gentle smile on your face and affirm, "**All the organs within my head region are healthy and balanced**."

Relax your entire body from head to toe.

With this technique, you will completely heal and relax your physical body to wash away all the toxic energies built up during the day. This Body Scan Healing technique, done daily right before you fall asleep, will allow you to completely relax and rejuvenates your body and mind for a good night's sleep.

For deep inbuilt negative emotions or weaknesses to be disintegrated, we practice with the help of the element of Fire.

Deep Cleansing to Heal the Mind – Burning Your Weakness

"Fire allows for rapid transformation. It provides the avenue to let go of the old story and drama, to transform, to renew and to be reborn."

- Dr. Alberto Villoldo

Practice the following technique at least once a week to give your mind a deep cleansing from negativity.

Take a plain sheet of paper. Sit in a quiet place and allow 15 minutes of your time for self-introspection.

First, write down "MY PHYSICAL WEAKNESS."

This means: what is it that is stopping you from working on your physical body? For example, if it is laziness that you feel is your weakness, then write down "my laziness." You will have multiple other weaknesses that are stopping you from achieving your physical goals. Write down all of those weaknesses.

In the same manner, write down "MY EMOTIONAL WEAKNESS".

What are the opposing forces that you must tackle mentally or emotionally?

Then write down "My weaknesses in my relationships".

What are the negative aspects of your behavior in your relationships towards your husband, wife, children, friends, etc.?

Finally, write down "MY PROFESSIONAL WEAKNESS".

Now BURN this chit of paper with the help of a candle flame. Burn it with the firm intention that, "I am burning all my weaknesses".

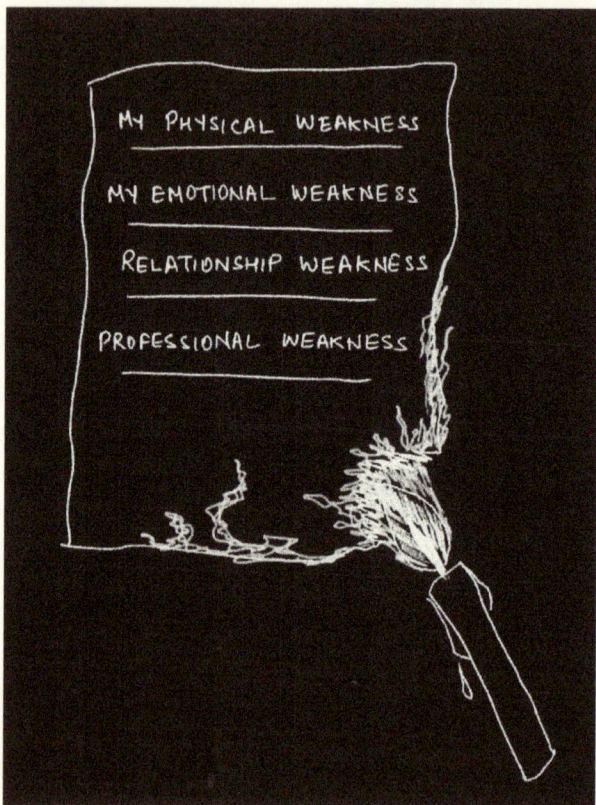

The fire element has a potent property to destroy and release you from the deep-rooted bonds of a negative quality that you may have either imbibed consciously or unconsciously through your experiences in your life. Let them go.

On another sheet, write down in bold letters

- MY PHYSICAL GOALS
- MY EMOTIONAL GOALS
- MY RELATIONSHIP GOALS
- MY PROFESSIONAL GOALS.

Put this up where you can see these goals in writing daily and work towards those goals daily.

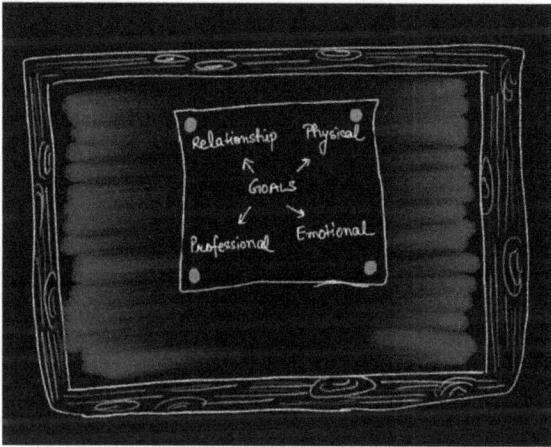

You can practice this whenever you feel low and running on low energies. This technique will make you feel lighter to move ahead positively in all aspects of your life.

You know yourself best! Choose to upgrade yourself.

Is There Evidence to Support Body Scan Healing Technique?

In the spiritual world, what we call the body scan technique is practiced as mindfulness-based stress reduction in medical practice. In a meta-analysis review done by Chen et al., the effect of mindfulness on sleep was studied. Mindfulness-based stress reduction (MBSR) was demonstrated to be a useful technique to help patients with insomnia. Those who practiced MBSR were shown to have improved sleep, and less depression and anxiety.[1] In a similar review of randomized controlled trials done in 2019, the short-term and long-term effects of MBSR were studied in patients with breast cancer. They concluded that practicing MBSR was able to bring down fatigue and reduce the stress and burden of the disease for at least six months to two years.[2]

Fire as an Element of Healing

Rituals are a part of our ancient culture and are being practiced in various ways across different cultures all over the globe. From the sacred yagnas that were performed by sages of yore to the Navajo chanting rituals of today, all contribute toward the healing of the self, others, and the environment. The element of fire is known to eliminate negative emotions and is deep-rooted in history as a major element of healing.[3,4]

In the Vedic era, sacred yagnas combined with the ceremonious chants of the sages would invoke energetic vibrations of the powerful presence of the Divine on Earth. It would also allow for healing from trauma and to connect people with the spiritual realm while destroying the negative boundaries that held them back.

Similarly, in today's world, fire is seen as a powerful source of healing. While the act of burning itself is dramatic – the ritual of doing so provides a sense of control to the person performing the ritual. As mentioned earlier on in this chapter – simply writing down negative emotions/thoughts on a piece of paper and burning it allows for a person to feel in control of their thoughts and to be able to destroy the negative ones, and paves the way for more positive emotions.

In an integrative review of the psychology of rituals, Hobson discusses in length the psychosocial impact of rituals.[5]

By creating a ritual for self-healing, one can block out anxious, intrusive thoughts, restore the feeling of order, regain agentic control and create a feeling of self-transcendence. In short, cleansing rituals can regulate emotions, performance goal state, and social connections to others.

Is There Science Behind Positive Affirmations?

Affirmations are positive statements that can help us challenge and overcome self-sabotage and negative thoughts.

Cascio et al.[6] identified reward center activation in response to self-affirmation practices. A simple affirmation like, "I will win so and so award" triggers the neural pathways and brings about changes to those areas of the brain that make us happy and positive.

So now that we know that it works, let us find out how. The human brain is innately lazy. It underwent evolution for the purpose of keeping us alive, for e.g., to make quick decisions about food and safety, and essentially not to make rigorous analytical assessments about everything. That means our brain is forever creating shortcuts. This introduces cognitive biases, which are beliefs that we all have without justification.

A few of these are listed below:

The Dunning-Krueger Effect: the tendency for novices to overestimate their skill and experts to underestimate their skill.

Observational Selection Bias: The tendency to notice more of something once it has been noticed, like when we buy a new car, and suddenly, we see that car everywhere.

The magic of affirmations is that they hijack these cognitive biases for our own benefit. They teach our brains a new way to think. If we repeat phrases or sentences that convince our brain, say I'll get the job or I will lose weight, I shall be happy no matter what, then our brain starts to subconsciously look for signs that can make this true.

Now which part of your brain actively functions for this reaction?

Meta-analyses across a variety of tasks have found that self-related processing is most often associated with increased activity in the medial prefrontal cortex (MPFC) and posterior cingulate cortex (PCC)[7,8]

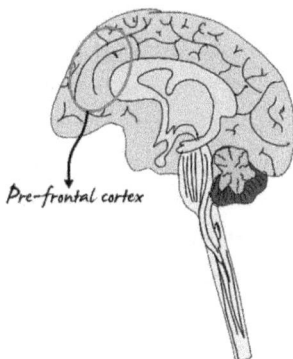

Pre-frontal cortex

It has been found that imagining the future with personally relevant, emotionally positive, and rewarding events is associated with changes in VMPFC, striatum, MPFC, and PCC.

Increased activity in the medial prefrontal cortex has also been shown to positively correlate with imagining positive (vs negative) future episodes[9].

The above neural associations highlight ways in which brain systems implicated in positive valuation and self-related processing may be reinforced by positive affirmations and suggest novel insight into the delicate balance of awareness and self-healing. They also suggest the role of future studies to examine the effects of self-affirmation interventions across a wide range of potential applications and outcomes.

Goals and Their Glory

"If you want to live a happy life, tie it to a goal; not to people or objects."

— Albert Einstein

Setting a goal is a value that is inculcated in us right from early childhood. A goal helps in defining purpose and motivates one to achieve success in that field. A goal could be related to physical health, mental wellness, social interactions, academics, professional growth, monetary, or overall personal development. One can have multiple goals at the same time based on one's core values and they need not be connected or achieved during the same time frame. However, creating and updating lists of goals keeps one motivated to work towards them.

The Agewell trial[10] for goal-setting behavior to promote physical and mental wellness among adults over 50 years of age studied three groups of participants – one where goals for physical and cognitive health were set, second where goals were set and mentored over the course of one year and the third being a control group who were educated about the importance of goal setting in a group discussion along with the other two groups. It came to light that the group that had set goals and was mentored regularly fared better than the other two groups in terms of physical and cognitive health proving that setting goals paved the way to create healthier lifestyles.

While setting a goal is simple enough, achieving them is not so easy. **Why?** This is because achieving a goal requires two things: a will and a way. In the true sense, a will is motivational – a drive to "want" to succeed while a way is more cognitive – the method in which the success is made achievable. Achieving a goal requires a change in behavior and being able to engage in that behaviour[11].

These behaviors could be simple-routine, complex-routine, simple- novel or complex-novel, each different from the other and requiring different degrees of motivation and skill. Lack of motivation (will) provides no way to achieve a set goal even if the task is simple or menial. A complex and new goal requires more motivation and skill than a simple and routine one would need. The dopaminergic "reward" system of the ventromedial prefrontal cortex of the brain allows for motivation to last in anticipation of the reward at the end of achieving a goal.

The different kinds of goals:

Complex and Routine , eg: Driving to work daily

Complex and New, eg: Driving in new city the first time

Simple and Routine, eg: Folding laundry

Simple and New, eg: Changing diaper for the first time

The "way" is more cognitive and is an intrinsic function of various areas in the brain including the lateral prefrontal cortex, temporal cortex, and the temporo-parietal junction. These in association with the dopaminergic systems work together to render executive functions resulting in the realization of a goal.

While it is important that we realize the importance of self-healing through the body scan healing technique, getting rid of negativity through the element of fire, and allowing positive affirmations and goals to define what we want from life, it is also imperative that we form a habit of them. It is not a one-time ritual that we should aim for, but rather a regular habit of cleansing our minds and our aura, changing our values and goals as life takes on a new shape at every stage, and working towards them step by step to achieve greater success and satisfaction of living well-rounded lives.

"I will either find a way, or I will make one."
— Phillip Sidney

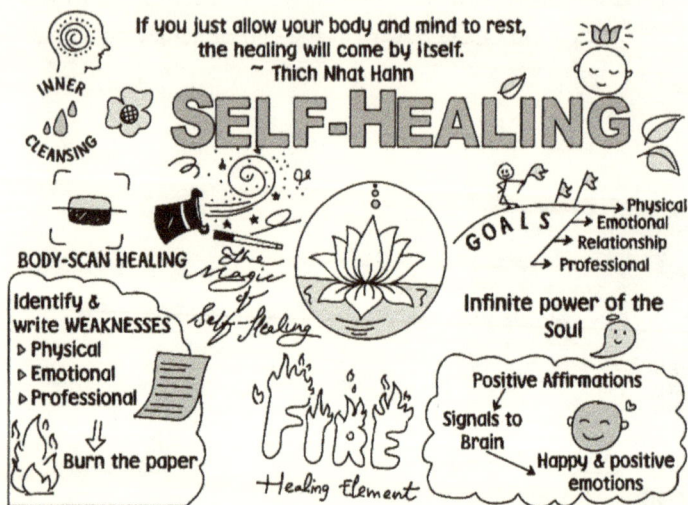

Would You Like to Read Further?

1. Chen TL, Chang SC, Hsieh HF, Huang CY, Chuang JH, Wang HH. Effects of mindfulness-based stress reduction on sleep quality and mental health for insomnia patients: A meta-analysis. J Psychosom Res. 2020 Aug;135

2. Schell LK, Monsef I, Wöckel A, Skoetz N. Mindfulness-based stress reduction for women diagnosed with breast cancer. Cochrane Database Syst Rev. 2019 Mar 27;3(3)

3. https://brewminate.com/the-ancient-practice-of-fire-rituals-for-healing-and-purification/

4. Kaptchuk TJ. Placebo studies and ritual theory: a comparative analysis of Navajo, acupuncture and biomedical healing. Philos Trans R Soc Lond B Biol Sci. 2011 Jun 27;366(1572):1849-58.

5. Hobson NM, Schroeder J, Risen JL, Xygalatas D, Inzlicht M. The Psychology of Rituals: An Integrative Review and Process-Based Framework. Pers Soc Psychol Rev. 2018 Aug;22(3):260-284.

6. Cascio CN, O'Donnell MB, Tinney FJ, Lieberman MD, Taylor SE, Strecher VJ, Falk EB. Self-affirmation activates brain systems associated with self-related processing and reward and is reinforced by future orientation. Soc Cogn Affect Neurosci. 2016 Apr;11(4):621-9. doi: 10.1093/scan/nsv136. Epub 2015 Nov 5. PMID: 26541373; PMCID: PMC4814782.

7. Northoff G, Heinzel A, de Greck M, Bermpohl F, Dobrowolny H, Panksepp J. Self-referential processing in our brain--a meta- analysis of imaging studies on the self. Neuroimage. 2006 May 15;31(1):440-57.   doi: 10.1016/j.neuroimage.2005.12.002.   Epub   2006 Feb 7. PMID: 16466680.

8. Denny BT, Kober H, Wager TD, Ochsner KN. A meta-analysis of functional neuroimaging studies of self- and other judgments reveals a spatial gradient for mentalizing in medial prefrontal cortex. J CognNeurosci.   2012   Aug;24(8):1742-52.   doi: 10.1162/jocn_a_00233. Epub 2012 Mar 27. PMID: 22452556; PMCID: PMC3806720.

9. D'Argembeau A, Xue G, Lu ZL, Van der Linden M, Bechara A. Neural correlates of envisioning emotional events in the near and far future. Neuroimage. 2008 Mar 1;40(1):398-407. doi: 10.1016/j.neuroimage.2007.11.025. Epub 2007 Dec 3. PMID: 18164213; PMCID: PMC2782689.

10. Nelis SM, Thom JM, Jones IR, Hindle JV, Clare L. Goal-setting to Promote a Healthier Lifestyle in Later Life: Qualitative Evaluation of the AgeWell Trial. Clin Gerontol. 2018 Jul-Sep;41(4):335-345

11. Berkman ET. The Neuroscience of Goals and Behavior Change. Consult Psychol J. 2018 Mar;70(1):28-44

9

HEALING THROUGH THE POWER OF BREATH

"When the breath wanders, the mind is also unsteady. But when the breath is calmed, the mind too, will be still."

— Hatha Yoga Pradipika

Just for a few seconds now, close your eyes with an aligned back and shift your awareness on your inhalation and exhalation for three minutes.

Now as you have opened your eyes and read these words, how do you feel? If you had done this sincerely, then you would have definitely sensed a subtle energy shift as you opened your eyes. In life, we need to 'pause' to shift our awareness by focusing on our breath to naturally move into 'Now', the present moment.

Our minds are always oscillating back and forth between the past and the future – this is the main reason why we slip into anxiousness. As we observe our breath, the mind is engaged with 'NOW'. Only when we pause we disconnect from the vicious cycle of thoughts and be able to 'LISTEN'.

The universe is always communicating with you. Our purpose is to silence ourselves to 'listen'.

Methodology

STEP 1 – TO MOVE TO THE PRESENT

Sit with your spine comfortably erect and aligned. Relax your shoulders. Cup your palms and place them over your knees with the palms facing up. If you are sitting on a chair, then make sure you have not crossed your legs and both soles of your feet are touched to the ground. Now close your eyes and gently shift your awareness to your breath. Notice the cool air moving in through your nostrils, and the warm air moving out. Make sure you are taking deeper and longer inhalations and exhalations. Notice and observe the subtle energy shift to the PRESENT MOMENT. Allow yourself to notice all your five senses: The air touching your skin, the fragrance in your room, and the sounds you hear. What do you see within when you have closed your eyes? Just notice... **What is your taste in life today?** Just observe and do not judge. Be non-judgemental to yourself and accept all, be it positive or negative. Do not force this … Just observe.

STEP 2 – AWARENESS AND CLEANSING OF OUR TOXIC ENERGIES WITH A CONSCIOUS ABDOMINAL BREATHING TECHNIQUE

Once we move to the present moment, we become extremely aware of how we feel about our body, mind, and emotion. This clarity is very much needed. Only when you can clearly pinpoint a problem can you find the solution to the problem?

The Process

In this process, we use our breath as a channel to discharge negative energies and imbibe positive energies.

Shift your awareness to the base of your spine. Any form of insecurities blocks this 'basic energy center.'

Focus on the tip of your spine and through your nostrils, deeply inhale and exhale through your pursed lips with the affirmation in your mind, "I release all the insecurities within my energy body." Here please note that you are practicing abdominal breathing exercises wherein as you inhale, the abdomen moves outwards, and as you exhale, the abdomen moves inward.

Feel lighter and lighter as you exhale deeply through your mouth. Conclude by inhaling deeply, affirming, "I am grounded, centered, and confident."

Now gently shift your awareness to your 'sacral energy center,' which resides three inches below your navel in your energy body. You can place your right palm over this energy center and deeply inhale and exhale through your mouth, affirming, "I relieve and let go of all the guilt."

Repeat the same till you feel extremely light at this energy center. Inhale deeply, as you affirm, "My sacral energy center is healed, and my creative energies flow freely through me."

Now gently shift your awareness to the spot three inches above your navel – the 'solar plexus energy center.' Place your palm here and exhale through your mouth, affirming in your mind, "I release and let go of all my stress, fear, and anger within me."

Inhale deeply through your nostrils and exhale through your mouth, repeating this affirmation till you feel very light at this energy center.

Inhale, as you affirm, "My solar plexus energy center is balanced, and I am filled with willpower to achieve my goals."

Shift your awareness gently to your heart center. Place your right palm on your heart center and inhale deeply and exhale out any form of hatred towards yourself or towards anybody.

Repeat the affirmation, "I let go of all the HATE." Repeat this as often as possible until you feel very light at your heart center. Now inhale deeply and affirm in your mind, "My heart is filled with love and compassion. My heart center is healed."

Shift your awareness gently to your throat center. Place your palm on your throat region.

Exhale deeply through your mouth with the affirmation, "I let go of all the blocks in my throat energy center". Keep exhaling till you feel lighter.

Now inhale deeply with the affirmation, "My throat energy center is balanced and filled with the power of communication and truth."

Shift your awareness toward the spot between your eyebrows.

Place your palm there and affirm with a deep exhalation, "I let go of all the illusions that block my third eye energy center."

Repeat till you feel the subtle energy shift at this powerful spot.

Inhale deeply and affirm, "I am filled with the power of intuition to SEE BEYOND, and I have clarity in every aspect of my life."

Shift your awareness towards the top of your head. Place your palm over your head, deeply exhale and affirm, "I let go of any form of attachment that blocks my crown energy center."

Repeat this as many times till you feel lighter, and then deeply inhale and affirm, "My crown energy center is healed and filled with the power to SURRENDER, and I have complete TRUST in my journey of life."

Now place both your palms over your knees with the palms facing upwards and sense this healing of your energy body that has taken place using your breath as your channel to heal.

Relax your entire body from head to toe.

With this technique, you will heal your energy body to wash away any form of toxic energies that were building up during the day.

"Regulate your breathing, thereby control your mind."
— B K S Iyengar

The Psychophysiology of Slow Breathing

While we all know that breath is life and that we live because we are breathing – but we are breathing passively. Whether we like it or not, till life is taken out of us, we are breathing, and that is the physical truth of it. **But what if we allow the breath to control us mentally?**

To be able to regulate our breathing can bring about an immense change in our mental and physical health and even change the way we look at life.

There have been numerous studies on the sacred *chakras* or energy centers in our body and the energies that emanate from them.

Christopher Chase describes the effects of *chakra* energy centers in the practice of acupuncture[1]. Seven *chakras* originate from acupoints on the physical body, and the complexity and energy levels of the chakras increase at each ascending level. It is important to note that the *chakras* penetrate the seven auric (energy) layers of the human biofields.[2] Each auric layer is an electromagnetic layer that surrounds the physical form of the human body, much like the lines of force surrounding a magnet.

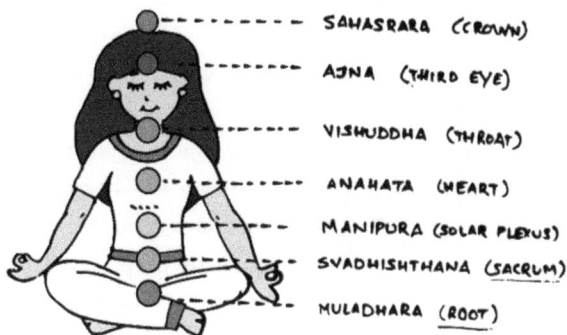

SAHASRARA (CROWN)

AJNA (THIRD EYE)

VISHUDDHA (THROAT)

ANAHATA (HEART)

MANIPURA (SOLAR PLEXUS)

SVADHISHTHANA (SACRUM)

MULADHARA (ROOT)

Each of the seven *chakras* has significance in our lives.[3]

The *Base* or *Root* (first) *chakra* is associated with the family, tribe, materialism, and abundance. The *Sacral* (second) *chakra* is associated with sexual and social fulfillment and emotional intelligence. The *Solar Plexus* (third) *chakra* is associated with career, capabilities, and self-esteem. The *Heart* (fourth) *chakra* is associated with romantic love and agape love. The *Throat* (fifth) *chakra* is associated with personal expression, self-awareness, and finding the truth. The *Brow* or *Anja* (sixth) *chakra* is associated with complex rational thought and insight.

The *Crown* (seventh) *chakra* is associated with surrender and recognition of the common ground of all living beings.

While some advocate the *chakra* system of acupuncture and other forms of natural healing, slow breathing techniques are also known to have psychophysiological outcomes.[4]

The breath is called *Prana*, which means both "breath" and "energy" (i.e., the conscious field that permeates the whole universe). "*Prana-Yama*" (literally, "the stop/control," but also "the rising/expansion of breath") is a set of breathing techniques that aims at directly and consciously regulating one or more parameters of respiration.

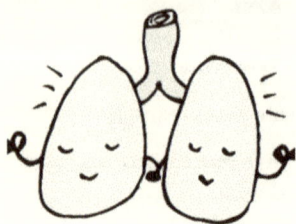

Slow breathing techniques have been shown to promote autonomic changes - increasing heart rate variability and respiratory sinus arrhythmia paralleled by Central Nervous System (CNS) activity modifications. Anatomically, functional MRIs highlight increased activity in cortical (e.g., prefrontal, motor, and parietal cortices) and subcortical (e.g., pons, thalamus, sub-parabrachial nucleus, periaqueductal gray, and hypothalamus) structures.

Psychological/behavioral outputs related to the mentioned changes are increased comfort, relaxation, pleasantness, vigor, and alertness, and reduced symptoms of arousal, anxiety, depression, anger, and confusion.

Paced breathing[5], or the Western culture promoted method of slow breathing, has been associated with relaxation and well-being and used for progressive relaxation therapy and autogenic training to guide our body's autonomic system.

One of the key elements of anxiety, panic attacks, or depression is heart rate variability which is the time interval between heartbeats.

Contrary to popular belief, this time interval is not necessarily a constant. Moods, time of day, the task at hand, etc., influence heart rate variability.

While psychotherapy can reduce the frequency of anxiety or panic attacks, they seldom cause a change in heart rate variability. But adopting methods of slow-paced breathing, yoga, and regular pranayama has been shown to regularize the variability in heart rate[6]. It can be controlled by ways of adopting slow breathing techniques.

The advantageous effects of yogic breathing on the neurocognitive, psychophysiological, respiratory, biochemical, and metabolic functions in healthy individuals have been studied extensively[7]. They were also found useful in the management of various clinical conditions. It is believed to reduce hypertension, reduce asthma symptoms, improve cardiac arrhythmias, reduce cancer-related fatigue, and better recovery from mental illnesses.

"Breath is the bridge that connects life to consciousness, which unites your body through your thoughts."

— Thich Nhat Hanh

When the breath wanders, the mind is also unsteady. But when the breath is calmed, the mind, too, will be still.
~ Hatha Yoga Pradipika

POWER OF BREATH

The Present Moment

"Now"

Slow Breathing

PAST   FUTURE

① Increased comfort
② Relaxation
③ Pleasantness
④ Vigor
⑤ Reduced anxiety

I am ... Grounded Centered Confident

BALANCING THE SEVEN CHAKRAS

Would You Like to Read Further?

1. Chase CR. The Geometry of Emotions: Using Chakra Acupuncture and 5-Phase Theory to Describe Personality Archetypes for Clinical Use. Med Acupuncture. 2018 Aug 1;30(4):167-178. doi: 10.1089/acu.2018.1288. PMID: 30147818; PMCID: PMC6106753.

2. McKusick ED. Tuning the Human Biofield. Rochester, VT: Healing Arts Press; 2014

3. Cross JR. Acupuncture and the Chakra Energy System: Treating the Cause of Disease. Berkeley: North Atlantic Books; 2008

4. Zaccaro A, Piarulli A, Laurino M, Garbella E, Menicucci D, Neri B, Gemignani A. How Breath-Control Can Change Your Life: A Systematic Review on Psycho-Physiological Correlates of Slow

Breathing. Front Hum Neurosci. 2018 Sep 7;12:353.

5. Jerath R., Crawford M. W., Barnes V. A., Harden K. (2015). Self-regulation of breathing as a primary treatment for anxiety. Appl. Psychophysiol. Biofeedback 40, 107–115. 10.1007/s10484-015-9279-8

6. Steffen PR, Bartlett D, Channell RM, Jackman K, Cressman M, Bills J, Pescatello M. Integrating Breathing Techniques Into Psychotherapy to Improve HRV: Which Approach Is Best? Front Psychol. 2021 Feb 15;12

7. Saoji AA, Raghavendra BR, Manjunath NK. Effects of yogic breath regulation: A narrative review of scientific evidence. J Ayurveda Integr Med. 2019 Jan-Mar;10(1):50-58. doi: 10.1016/j.jaim.2017.07.008. Epub 2018 Feb 1. PMID: 29395894; PMCID: PMC6470305.

10

Grounding

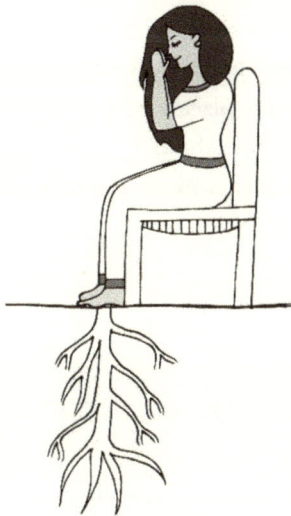

"Flying starts from the ground. The more grounded you are, the higher you fly."

-J.R. Rim

Grounding is the physical connection between the electrical frequencies of the human body and the Earth. Practicing grounded meditation allows you to put your mind in the present and learn to feel more balanced and aware. You might feel uneasy, restless, and uncomfortable when your mind is unfocused. Grounding helps you feel calm, peaceful, and centered. After meditation, grounding yourself will bring balance to the restoration of your energies. You stabilize your energies in such a way that you feel uplifted and, at the same time, rooted.

Methodology

STEP 1- Prepare yourself

- Turn off all electronics, including cell phones and televisions, which can cause distractions.
- Find a quiet place where you can be alone for 20 minutes. Place both your feet on the ground. If possible, find a place with grass or soil on which you can stand with your bare.
- You can stand or sit on a chair and place your feet firmly on the ground or sit directly.
- Close your eyes.
- Take a moment to make sure your body is relaxed and comfortable.

STEP 2 - Focus on your breathing

- Close your eyes and focus on your breathing

- Notice your thoughts or any distractions, push them away slowly and, return your awareness to breathing.

Notice your thoughts or any distractions, push them away slowly and, return your awareness to breathing.

STEP 3 - Begin the Visualisation Method 1: Tree Meditation

- Visualize yourself sitting on a tree trunk.
- Focus on the energy in your body flowing down to the Earth. The tree shall become an extension of your body, extending through your feet.
- Continue to imagine this tree trunk traveling down through the Earth until it finally reaches the center of the Earth.

- While you are breathing, let every negative feeling escape your body. Leave all feelings of pain, frustration, anger, or bitterness in the center of the Earth.
- Imagine Earth's energy flowing back up through you. Push this energy upwards.

STEP 4 - Come Back to Stillness

- Before opening your eyes, rest in a state of centered and calm for a few moments.
- When you feel ready, bring your mind back to your body.

The more you practice grounding meditation, the more natural it becomes.

By taking a few moments each day to practice this, you'll not only experience better energy in your daily life but will also help you re-focus. You can thus find your inner calm and peace no matter where you are.

Scientific Basis of Grounding

Humans, with time, have evolved with amazing body structure advancements and adapted to changing environments. However, with the evolution of the human body came the progress of technology and materialistic advancement. Though these changes led to some wonderful innovations that we have been using in our day-to-day lives, racing through newer advancements has left behind some older ways of life that were proven to be both physically as well as mentally sound for our health.

One such lifestyle that has been left behind is walking barefoot or sleeping on the ground. In the past, humans used to walk barefoot on the Earth or with leather-soled shoes or moccasins that were electrically conductive. Also, the habit of sleeping directly on the ground or over electrically conductive materials has been found to be clinically beneficial. Given the advent of shoes made of non-conductive materials like plastic and rubber or beds made of insulating materials, these habits are now almost obsolete.

In the past, humans used to walk barefoot on the Earth or with leather-soled shoes or moccasins that were electrically conductive.

Earthing researchers have debated over a new deficiency disorder that can be termed as "electron deficiency" caused by lack of grounding or earthing, i.e., surface contact with the Earth.[1] These electrons are essential for the normal functioning of our immune systems, and a deficiency of the same can cause a wide variety of health implications. Electrons are nature's original and best antioxidants. Most dietary antioxidants act by delivering electrons, which neutralize the ROS (reactive oxygen species) responsible for inflammation.[1] Ober described an increase in the occurrence of inflammatory disorders with insulating shoes and found an increased incidence of diabetes associated with increasing sales of athletic shoes with insulating soles.[2] Lack of these electrons may be responsible for several chronic ailments like autoimmune disorders, allergies, or even cancers.[1,2]

One of the reasons that earthing works quickly on inflammation anywhere in the body is a continuous system of "biological wires" composed of semiconductor proteins forming a system called "the living matrix".[3] Maurice et al., in their study on "The Biologic Effects of Grounding the Human Body During Sleep as Measured by Cortisol Levels and Subjective Reporting of Sleep, Pain, and Stress," concluded that there was normalization of circadian cortisol levels after eight weeks of grounding in patients, who suffered from sleep dysfunction, chronic pain and stress.[4] These endocrine effects were later confirmed independently by Karol and Pawel Sokal, who found the biological effects of earthing on blood chemistry during sleep and also commented that continuous earthing of diabetic patients manifested a reduction in blood glucose levels.[5]

Ober, in 2006 also studied the effects of grounding on reductions in overall stress levels and tensions by causing muscle relaxation and shifting from stressful sympathetic activation to a calming parasympathetic state.[1,2] Other biological benefits reported in the literature include accelerated healing of injuries, reduction in blood viscosity, improved blood flow, decreased hormonal and menstrual symptoms, reduction in blood pressure in hypertensive patients, and increased heart rate variability, a measure of the health of the cardiovascular system.[6]

Additionally, it was observed that earthing could prevent and treat several respiratory disorders like bronchial asthma and Chronic Obstructive Pulmonary Disease (COPD).[1] Recently, it was also found to prevent and treat serious COVID-19 infections.[7,8] Chevalier et al. experimented with this by placing grounding patches over each lung, with the wires plugged into the ground port of the electrical system, and commented that earthing leads to relief of cytokine storm and deadly pneumonia occurring as a result of COVID-19 infections.[1]

Hence, practicing 'grounding' can help boost the body's functioning and can increase life expectancy by improving the sleep cycle, lowering stress, and reducing the occurrence of chronic ailments. Moreover, it can help delay the aging process – by preventing mitochondrial "self-destruction" by ROS produced by oxidative metabolism via the delivery of antioxidant electrons to the mitochondrial matrix.[9]

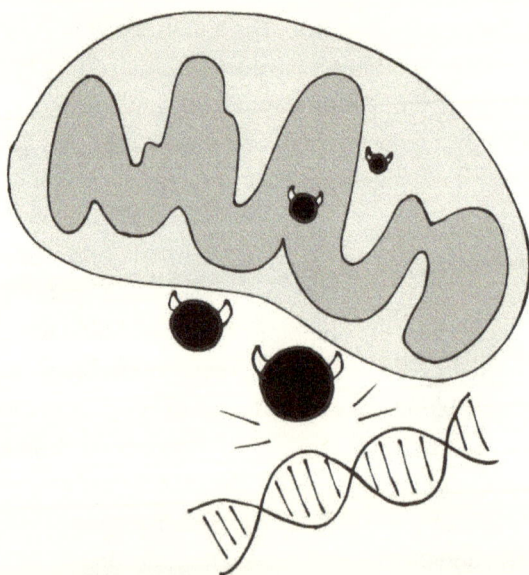

Mitochondrial "self-destruction" by ROS produced by oxidative metabolism contributes to the aging process, which can be delayed through the practice of grounding.

The practice of grounding meditation lends literal meaning to the process of 'grounding yourself' or 'staying grounded' – phrases that we commonly use to describe the process of anchoring oneself to reality, staying true to one's roots, being humble, strengthening oneself, and having a firm foundation. Regular practice of this meditation may help not only in strengthening oneself mentally, but clinical research suggests that it may help one prevent a myriad of systemic illnesses and also help one develop a robust constitution so as to delay aging and promote vitality.

Would you like to read further?

1. Oschman JL. Illnesses in technologically advanced societies due to lack of grounding (earthing). Biomed J. 2022; 3:S2319- 4170(22)00152-4.

2. Ober C. Grounding the human body to neutralize bioelectrical stress from static electricity and EMFs. ESD Journal. [Internet]. 2000 [cited 2022 Nov 20]; Available from: http://www.esdjournal.com/articles/cober/ground.htm

3. Friesen DE, Craddock TJ, Kalra AP, Tuszynski JA. Biological wires, communication systems, and implications for disease. Biosystems. 2015;127:14-27.

4. Ghaly M, Teplitz D. The biologic effects of grounding the human body during sleep as measured by cortisol levels and subjective reporting of sleep, pain, and stress. Journal of Alternative and Complementary Medicine. 2004;10(5):767–76.

5. Sokal K, Sokal P. Earthing the human body influences physiologic processes. J Altern Complement Med. 2011;17(4):301-8.

6. Chevalier G, Sinatra ST, Oschman JL, Sokal K, Sokal P. Review article: Earthing: health implications of

reconnecting the human body to the Earth's surface electrons. J Environ Public Health. 2012;2012:291541.

7. Abdul-Lateef Mousa H. Prevention and treatment of COVID-19 infection by earthing. Biomed J. 2022; 17:S2319-4170(22)00121-4.

8. Ober C, Oschman JL (2020) Prevention and/or recovery from coronavirus infections. Int J Clin Endocrinol Metab 6(1): 022-024.

9. Ober, C., Sinatra, S. T., Zucker, M. (2014). Earthing: the most important health discovery ever! Second edition. Laguna Beach, CA, Basic Health Publications

THE CONCLUSION

Your journey starts now

"Until you make the unconscious conscious, it will direct your life, and you will call it fate."
— Carl Jung.

We have come to the end of this book. But we need to understand that "Every end is a new beginning."

Let's practice these techniques sincerely to experience the "MAGIC" in our lives. Let us transcend negative emotions to balance our body, mind, and spirit. May these theories, concepts, pieces of evidence, and research help imbibe more faith in our practices.

Let us grow in our journey to bring out our inner beauty and enhance our lifestyle. Let's strive to live our lives to the fullest of our capacity with JOY, LOVE, and BLISS.

The mantra to reinforce here is "ENERGIES WILL WORK FOR ME LIKE MAGIC ONLY WHEN I HAVE COMPLETE FAITH IN IT."

You are a special part of my journey now — let's together take this forward to spread knowledge and wisdom to bring more smiles into our world.

Thank you!

www.ingramcontent.com/pod-product-compliance
Lightning Source LLC
Chambersburg PA
CBHW030457100426
42813CB00002B/251